Heritage, Crafting Communiti and Urban Transformation

I0037783

This book emphasises the need to empower marginalised communities to contribute to decision-making processes within policy realms. It contributes to ongoing debates in the social sciences about infrastructure rights and citizenship, and it throws insight on human–infrastructure interactions in the informal neighbourhoods of the global South.

The book delves into the complexities of caste, gender, class, and political identities and affiliations associated with the multiple factors of inclusion and exclusion particularly in the case of access to infrastructure in informal settlements in urban areas with an added productive function. This book is about how this historic inner-city, situated, religious idol-crafting community is transforming due to factors including access to physical and social infrastructure, local governance policies, sociopolitical hierarchies, and complexities of informal tenure. Drawing on sociocultural norms and values of idol-crafting practices, it documents, analyses, and presents the networks and relations of the neighbourhood through a spatial and material lens. Findings contribute to understanding how traditional practices of a crafting community are adapting, appropriating, producing, and reshaping informal spaces in Kumartuli.

The book is aimed at academic audiences across the world researching creative industries, Kolkata's regeneration agenda, and cultural tourism. It will be of interest to the wide disciplines of Urban Studies, Development Studies, Architecture and Planning, and Culture and Tourism Studies.

Debapriya Chakrabarti is a researcher in the field of urban studies at the Manchester Institute of Innovation Research and teaches at the Manchester School of Architecture. She is trained as an architect and urban planner. Her research interests lie at the intersection of urban regeneration, cultural industries, and place-based development policies.

Routledge Studies in Urbanism and the City

Lahore in the 21st century
The Functioning and Development of a Megacity in the Global South
Mohammad A. Qadeer

Writing the City Square
On the history and the histories of city squares
Martin Zerlang

The Anatomy of Inclusive Cities
Insight into Migrants in Selected Capital Cities of Southern Africa
Hangwelani Hope Magidimisha-Chipungu and Lovemore Chipungu

Reflecting on the City Through Literature
Urban Spaces, Differences and Embodiments
Daan Wesselman

Splintering Towers of Babel
Paradoxical Architectures and Urban Infrastructures
Liora Bigon and Edna Langenthal

Urban Ethics as Research Agenda
Outlooks and Tensions on Multidisciplinary Debates
Raúl Acosta, Eveline Dürr, Moritz Ege, Ursula Prutsch, Clemens van Loyen and Gordon Winder

Heritage, Crafting Communities and Urban Transformation
Durga Puja Festival, Kolkata
Debapriya Chakrabarti

For more information about this series, please visit www.routledge.com/Routledge-Studies-in-Urbanism-and-the-City/book-series/RSUC

Heritage, Crafting Communities and Urban Transformation
Durga Puja Festival, Kolkata

Debapriya Chakrabarti

Routledge
Taylor & Francis Group

LONDON AND NEW YORK

First published 2024
by Routledge
4 Park Square, Milton Park, Abingdon, Oxon OX14 4RN

and by Routledge
605 Third Avenue, New York, NY 10158

Routledge is an imprint of the Taylor & Francis Group, an informa business

© 2024 Debapriya Chakrabarti

The right of Debapriya Chakrabarti to be identified as author of this work
has been asserted in accordance with sections 77 and 78 of the Copyright,
Designs and Patents Act 1988.

All rights reserved. No part of this book may be reprinted or reproduced or
utilised in any form or by any electronic, mechanical, or other means, now
known or hereafter invented, including photocopying and recording, or in
any information storage or retrieval system, without permission in writing
from the publishers.

Trademark notice: Product or corporate names may be trademarks or
registered trademarks, and are used only for identification and explanation
without intent to infringe.

British Library Cataloguing-in-Publication Data
A catalogue record for this book is available from the British Library

ISBN: 978-1-032-37647-9 (hbk)
ISBN: 978-1-032-37650-9 (pbk)
ISBN: 978-1-003-34122-2 (ebk)

DOI: 10.4324/9781003341222

Typeset in Times New Roman
by Apex CoVantage, LLC

Dedicated to my grandmothers,
Ramarani Chakrabarti and Nilima Biswas,
and the artists of Kumartuli.

Contents

Figures

Tables

Preface

Autumnal Durga Puja is an annual Hindu religious and cultural festival, exhibiting a series of celebratory activities around hand-crafted idols and decorative pavilions that occupy Kolkata's streets for approximately ten days in September–October, attracting millions of visitors. *Puja* literally means worship, and Durga Puja means the worship of the Hindu deity Durga along with her mythological children, Lakshmi, Saraswati, Ganesh, and Karthik, during the autumn. Like all other Hindu religious worships, deities are crafted from natural materials, ritually worshipped, and immersed into waterbodies for the material to disintegrate: a sustainable cyclic process. Now more than 300 years old, the incorporation of Kolkata's Durga Puja in the 2021 UNESCO Intangible Cultural Heritage of Humanity (ICH) list is a tribute to the significance of the festival and the communities associated with crafting its exceptional idols and makeshift pavilions (*pandals*). This book presents the historical and contemporary socio-spatial arrangement of one community and neighbourhood profoundly associated with the heritage of Durga Puja in Kolkata. In Kolkata, the religious idols are sculpted of straw and clay in Kumartuli. This historic caste-based neighbourhood is increasingly being threatened by the city's urban redevelopment agendas, growing consumption, and mounting tourism pressures, making this book exceptionally relevant for local policymakers and international stakeholders, including UNESCO.

The city's cultural legacy led to the current cultural industries' policies and the nomination of the festival of Durga Puja to UNESCO's Intangible Cultural Heritage list by the current local government. Durga Puja, I argue, lies at the intersection of religious, social, and cultural politics that resulted from the complex nationalist struggle and the post-independence neighbourhood club and political party grassroots organisational foundations in urban Bengal. A seemingly domestic, feminised culture of worshipping the religious deity, over 300 years of urban Bengal's history, became an icon to showcase wealth, power, and the changing political landscape through sponsored immersion rallies by the state and the nomination and inscription of the festival as Intangible Cultural Heritage of Humanity.

The festivities have grown since the late nineteenth century, peaking extensive public participation with consistently increasing numbers of idols crafted over the years in the idol-making quarter. This has brought about issues regarding the accommodation of the growing production and storage requirements in the already-crowded neighbourhood with the deteriorating physical environment. Seeking to respond to this situation, the government proposed a redevelopment programme to 'improve' the area's spatial character, but this, as revealed in this book, was a failure. As always, the neighbourhood remains under pressure to serve the growing demand while transformations are underway. For instance, some idol-makers now occupy workshops outside Kumartuli and engage with suppliers and buyers in increasingly extended supply chains and global markets. As a result, there has been tension between the territorial practices of local government and the more relational spaces adopted by those living and working in the area.

This book is about how this historic inner-city, situated, religious idol-crafting community in India is undergoing continuous socio-spatial transformation due to factors including access to physical and social infrastructure, local governance policies, socio-political hierarchies, and complexities of informal tenure. The book combines findings from my doctoral thesis and postdoctoral fellowship fieldwork to understand how traditional practices of a crafting community are adapting, producing, and reshaping informal spaces of the neighbourhood. This book critically examines the aftermath of the stalled Kumartuli Redevelopment Plan proposed by the Kolkata Metropolitan Development Authority (KMDA), the associated complexities of the ownership and tenure status, and the increase in public grievances associated with the collective resistance to the plan.

The city of Kolkata has been my training ground as an architect, and following the footsteps of many inspiring seniors in my profession, I went on to complete a master's degree in City Planning. Crafts and cultural industries based in an inner-city location like the Kumartuli neighbourhood and the ongoing changes there during my studies always interested me: first as an undergraduate student of Architecture and then as a postgraduate student searching through probable dissertation ideas. I had been curious of the spatiality and the practices associated with cultural crafts such as idol-making and the allied *shola* and garments industry. However, I struggled to find the *critical question* that connects these embedded practices and ties them to the social and cultural heritage of not only the festival but also the place. I started trying to unravel idol-crafting's political economy based in Kumartuli. I followed the crafting process, materiality, and the practices associated with idol-making and found how these are intertwined with the community and craftspeople of Kumartuli. This monograph is a combination of years of research and quest. Not only do I draw from my learnings as a humanities and social sciences researcher and my training as an architect, but also from my lived experiences as a born and bred Bengali from Kolkata.

In addition, I also use a range of diverse voices in the form of participatory photo voice and a deliberative workshop of this crafting community which I felt were not consulted during the earlier redevelopment planning process like in any other similar redevelopment activity. The reflections include particular insights on existing infrastructures, public services, and the politico-spatial discomfort and grievances of the communities of Kumartuli. I mostly took photographs in 2017, 2018, and 2022. However, a range of photographs taken by my research participants based in Kumartuli as part of the participatory photo-study has also been presented. These photographs are annotated as an alpha-numeric index to denote the participant and corresponding photograph number. I have explained the details of this photo-study in the methodological appendix.

My personal associations with Durga Puja are manifold. I was born and raised in Kolkata, where people not only look forward to Durga Puja but also plan their year around the autumnal festivities. So, despite following the Gregorian calendar for official purposes, Bengali households in Kolkata have a Bengali calendar that starts with New Year's Day in April. Moreover, I remember that the first thing we looked up on the calendar was the dates of Durga Puja, which usually starts at the end of September or early October. Schools have autumn breaks for the festival. Hence, it is fair to say that Durga Puja is profoundly embedded within the city of Kolkata's history, culture, and socio-political legacy that ultimately culminated in the faith-led politicised immersion rallies and the decision of getting nominated to UNESCO's ICH list.

However, before delving deeper into the religio-political entanglements, I should narrate Kolkata's (or Calcutta's) brief economic and political history to contextualise Durga Puja. The erstwhile Calcutta, the British colonial capital of India, at the beginning of the twentieth century transformed into the present Kolkata, while suddenly being termed the 'dying city' in the closing decades of the twentieth century in the words of India's then Prime Minister Rajiv Gandhi (Bumiller 1985; Lewontin 1985; Datta 1992). Within a century, the city had witnessed an array of remarkable political events, starting from the shift of the colonial capital to Delhi (c. 1911), the independence and the partition of India resulting in mass migration (c. 1946–50 and c. 1971), experiencing the post-independence government of the socialist democratic Congress Party (c. 1950–76), and the coming to power of a democratically elected Left Front government, led by the CPIM (c. 1977) after massive political unrest. All these events significantly shaped public opinions towards governance and impacted the city's growth and development.

It is imperative to note the dates mentioned so far to discuss the 'dying city' discourse because the political history of Bengal is somewhat related to the changes in the history of settlement in present Kolkata and its demographical condition. Kolkata has witnessed one of the most massive migrations through the partition of Bengal and the subsequent refugee influx from then East Pakistan, which is now Bangladesh (c. 1905, 1942, 1948–51, and Bangladesh Liberation War in 1971). The partition of Bengal also triggered the decline of one of the most significant agro-based industrial sectors: jute. The effect of the decline of the industries and the economic stagnation meant that Bengal, specifically Kolkata's suburbs, faced a diminished growth rate in the late 1990s and 2000s, much slower than Kolkata's peer million-plus cities of India (Shaw and Satish 2007). However, I argue here that Kolkata was faced with waves of mass migration post-independence while it was still tackling the trauma of partition and was completely lacking the infrastructure and housing facilities to accommodate the influx at a rate much higher than other Indian cities. Many refugee colonies and settlements were planned to tackle the crisis and to provide for the influx.

Furthermore, when India was preparing for economic liberalisation and opening to the global free market economy, the urban agglomeration around Kolkata was busy addressing the economic slowdown. The unending crises of resources and funding in the various private and public sector medium and small enterprises (jute mills, paper mills, dairy and poultry industries, and the like) resulted in labour strikes and subsequent closures in a shifting political scenario and crippling infrastructure. The 'dying city' allegory was accompanied by differential policies of the central and state governments due to the different political parties in power. The federal state government and the local state government have always been opposition parties; hence, allocating resources has always been a reason for public discontentment.

After the country's economic liberalisation and globalisation, Kolkata started to reshape with the advent of the information technology (IT) and services sector on its fringes. The Left Front Government faced adverse reactions and massive public agitations due to the plan of developing a controversial Tata Motors car manufacturing plant in 2007 on a fertile patch of arable land in the Gangetic plains. This agitation subsequently led to the power shift from the Left Front to Mamata Banerjee-led Trinamool Congress in 2011 (Banerjee 2012; Dutt 2018). While the Left Front government's policies have been controversial and always under scrutiny since the Indian economic liberalisation of 1991 (Roy 2004), it is also to be noted that the coming to power of the Trinamool Congress (hereafter TMC), an offshoot of the Indian National Congress, furthered the market-driven neoliberal urban policies in West Bengal (Bose 2013). In fact, the last term of the Left Front government was dotted with public agitations and politically motivated collective resistance to large governmental 'development' and 'redevelopment' projects. The failure of the Kumartuli redevelopment plan was one such project. Since the TMC came to power, policies directed towards promoting religious events, such as Durga Puja and other festivals, were prioritised. These state-sanctioned festivities represent a religio-political agenda that brings the cultural heritage of the festival to the forefront and, as revealed through this book, might marginalise a large section of the crafting community involved in preparing the various artefacts for the festival.

Besides being popularly known as the 'City of Joy' (Lapierre 1985), Kolkata continues to be known as the 'Cultural Capital of India' (Roy and Raychaudhury 1974, Bradbury and Sen 2020). The Nobel Laureate Abhijeet V. Banerjee, while talking about Kolkata in a nostalgic article, points out that the culture was historically the binding element in terms of nurturing creativity and talents in the arts and literature that once earned international attention. Talents were attracted to come to Kolkata in order to thrive (Banerjee 2011). The celebration of Durga Puja in Kolkata, besides the religious rituals, reflects how the city's residents interact with popular culture and how music, food, and social meetings at cafes and tea houses are reflections of the cultural vibe of Kolkata. British Council estimates the several businesses the Durga Puja triggers that have shaped cultural assimilation in Kolkata, including the publication of books and annual magazines, music and theatre, cinema and fashion, and much more, contributing millions of rupees to the economy (British Council 2019).

The city used to be the seat of the literary and cultural revival in eastern India during the colonial period and beyond (Banerjee 2011). Festivals, an integral part of India's urban landscape, boost social and cultural exchanges. However, despite its grand festivals and vibrant popular culture, Kolkata continues to be regarded as the city that once was—the 'had been' city or even the 'dying' city (Bumiller 1985). Chattopadhyay (2005), in a detailed ethnographic study of the city, addresses the common misrepresentation of Kolkata as a 'metaphor of urban disaster' by the nineteenth-century British discourse that problematises poverty as the driving force for many of the urban problems faced in the innumerable slums of the city since its inception. On the contrary, the image of Kolkata has always been that of the residences of the *bhadralok*, bestowed by the power of the *babu* (landed gentry and rich businessmen) culture. While Kolkata's urban dream continued to be shaped by the upper classes, the poorer, marginalised communities, remained in shanty-like settlements in the fringes, rather underrepresented in popular culture or the public imagination.

The civic authority of the city of Kolkata is the Kolkata Municipal Corporation (KMC), the oldest urban local body in the metropolitan area of Kolkata, founded in 1726. It comprises 144 wards encompassing an area of 205 sq. km as of 2016 (KMC 2017). The area of KMC has increased over the years, mostly towards the southern and southwestern fringes of the city. Wards 1 to 100 have existed since 1965 and would unofficially be regarded as the core (inner-city), and wards 101 to 144, added after 1984, constitute the peripheral wards displaying high inequalities in density and condition of building stock (Shaw 2015). A master plan for the Kolkata Metropolitan Area (KMA) was developed in 1966 by the Calcutta Metropolitan Planning Organization (CMPO). The Kolkata Metropolitan Development Authority (KMDA), the successor of CMPO and founded in 1970, is an exclusive agency for planning and project execution in KMA. The demographics of West Bengal show that about 51 per cent of the urban population of the whole state are residents of the KMA, which comprises primarily urban areas and some rural areas; about 30 per cent of the population of KMA live within the area of Kolkata Municipal Corporation which predominantly is urban (KMC), and only about 8.7 per cent of the population live in about 40 per cent of KMA's rural land area (Shaw 2015). The civic authorities have been referenced repeatedly in this book due to their involvement in Kumartuli directly and indirectly in many ways on an everyday basis.

References

Banerjee, A. (2011). Once there was a city—India—Hindustan Times. *Hindustan Times*. Kolkata.

Banerjee, S. (2012). Post-election blues in West Bengal. *Economic and Political Weekly*. **47**: 10–13.

Bose, P.S. (2013). Bourgeois environmentalism, leftist development and neoliberal urbanism in the City of Joy. *Locating Right to the City in the Global South*. London, Routledge: 127–151.

Bradbury, J. and A. Sen (2020). Introduction: Calcutta characters. *Contemporary South Asia*. **28**: 427–433.

British Council (2019). Mapping the creative economy around Durga Puja 2019. British Council.

Bumiller, E. (1985). As Calcutta lies dying, beauty and squalor embrace. *The Washington Post*.

Chattopadhyay, S. (2005). *Representing Calcutta: Modernity, Nationalism and the Colonial Uncanny*. London and New York, Routledge.

Datta, P. (1992). Review essay: Celebrating Calcutta. *Urban History*. **19**: 84–98.

Dutt, I.A. (2018). Ten years after Tata Motors exit, Singur in West Bengal still a wasteland | Business Standard News. *Business Standard*. Singurs.

KMC (2017). Official website of Kolkata Municipal Corporation.

Lapierre, D. (1985). *The City of Joy*. Translated by Kathryn Spink. Garden City, NY, Doubleday.

Lewontin, M. (1985). Calcutta: City of contrasts where poverty does not diminish pride. *The Christian Science Monitor*.

Roy, A. (2004). The gentleman's city. *Urban Informality: Transnational Perspectives From the Middle East, Latin America, and South Asia*. A. Roy and N. AlSayyad. Lanham, MD, Lexington Books: 147–170.

Roy, P. and P. Raychaudhury (1974). Need for a municipal library system for the city of Calcutta. *International Library Review*. **6**(1): 43–49.

Shaw, A. (2015). Inner-city and outer-city neighbourhoods in Kolkata: Their changing dynamics post liberalization. *Environment and Urbanization Asia*. **6**: 139–153.

Shaw, A. and M. Satish (2007). Metropolitan restructuring in post-liberalized India: Separating the global and the local. *Cities*. **24**(2): 148–163.

Acknowledgements

Turning seven years of doctoral and postdoctoral research into a book would not have been possible without the kindness of my friends, mentors, colleagues, and family. Each and every one in this journey has, in their own way, contributed towards enriching my life and career. I am sincerely grateful to Deljana Iossifova and Graham Haughton at The University of Manchester not only for encouraging me in my academic and intellectual pursuit during my dissertation and beyond but also for their friendly advice, guidance, help, support, and feedback. This book would not materialise without Deljana's generous and inspiring words of enthusiasm through my ups and downs. I am immensely grateful for the support of my mentor, Stephen Walker, during my postdoctoral fellowship, who has, time and again, read and commented on my doctoral thesis and parts of this manuscript. He has offered practical guidance and endless encouragement to finish this book. I thank Swati Chattopadhyay for her critical feedback and supportive comments during my doctoral examination and afterwards. I also thank academic members of the Department of Architecture, The University of Manchester, Leandro Minuchin and Lukasz Stanek, for constructive comments, academic advice, and feedback during the dissertation stage of this research.

My SusInfra and Urban Infrastructural Reconfigurations reading group colleagues Ulysses Sengupta, Stelios Zavos, Leon Felipe Tellez Contreras, Youcao Ren, Purva Dewoolkar, Junyan Ye, Dongyang Mi, Srijon Barua, Subham Mukherjee, Qiwei Peng, Ziqui Ren, Elsa Holm, Mark Stanov, Rujin Wang, Rati Chaudhary, and Xin Li have all shared enriching discussions and critical perspectives that helped shape my arguments. I feel fortunate to have found a truly supportive network of friends. Each one of our long conversations around books, chapters, articles, dissertations, conferences, food, and life, in general, has always been pleasurable and heart-warming. My colleagues at Manchester Institute of Innovation Research and Manchester Urban Institute, Elvira Uyarra, Kieron Flanagan, James Evans, Joe Ravertz, Alina Kadyrova, Mabel Sánchez Barrioluengo, Huma Javaid, Aarti Krishnan, Debbie Cox, Holly Crossley, and Wendy Walker have provided support and guidance when needed. My colleagues at MSA, Lucy Montague, Eamonn Caniffe, Matthew Steele, and Michael Coates, have been affectionate, considerate, and friendly. This research has benefited from discussions with colleagues Diana Mitlin, Nicola Banks, Seth Schindler, Ezana Haddis, Tom Gillespie,

Khalid Nadvi, Maria Rusca, Melanie Lombard, Philipp Horn, and Sally Cawood at various points.

I appreciate the generous funding from The University of Manchester that I received during different stages of this research and the UKRI Economic and Social Research Council (ESRC) Postdoctoral Fellowship programme, without which the conception and completion of this monograph would not have been possible. I am thankful for the resources and support service teams at The University of Manchester, specifically the library and IT services that have provided continued academic and technological support.

My editors at Routledge, Faye Leerink and Prachi Priyanka, have been immensely supportive and patient. I thank Faye and Prachi for their continued support through the book proposal to the completed manuscript. I have received very generous proofreading support from family and friends—Dishari Chakrabarti, Sayantan Das, Krishna Biswas, and Debasmita Shashmal. I owe immensely to each of them. My deepest gratitude to my Kolkata participants—the artists and residents of Kumartuli who let me into workshops and homes and generously gave up their time for this research. I thank Kolkata colleagues in architecture, planning, academia, and municipal services, including Konaditya Bhattacharya, Sourav Sen, Amitava Sengupta, Suparna Dasgupta, Aurobinda Debnath, Sandipan Chatterjee, and Tapan Dhar. I also thank participants in the Deliberative workshop—Mahalaya Chatterjee, Debabrata Ghosh, Joy Sen. I express my gratitude also to the Kolkata Municipal Corporation, particularly to commissioners Khalil Ahmed and Vinod Kumar and joint secretary Tapas Chowdhury,

I sincerely thank all my friends and family who assisted me in this research by taking photographs, notes, and drawings during and beyond my fieldwork, including Nabarun, Rohan, Tamonash, Tatai, Soumya, Tushar, Debjoy, Smarajit, and many more. They have all provided me with valuable data in their own way and helped shape this research. My friends, old and new, but in no particular order—Arunima Kisku, Gargi and Jaydip Neogi, Rashmi Guharay, Mouli Majumdar, and Dibyadyuti Roy—have all showered me with love and kindness time and again. They have provided me with much-needed support and nourishment through food, drinks, intellect, and gossip throughout the last seven years of my research journey, both in the UK and in Kolkata.

I finished writing this book after the COVID-19 pandemic and sat at home. I am lucky that I have not had technology issues; my laptop and printer cooperated with me. I appreciate the transport workers in the rail and aviation industry who continued to serve humanity during and after the pandemic. It would not be possible without their brave and consistent support that allowed me to take flights and trains back and forth to complete the remainder of my research to finish this book. I would also like to express my gratitude to poets and artists, particularly musicians from South Asia, for creating the diverse wealth of magical songs that provided me with solace during long hours of lonely writing. I want to thank a few very supportive people in my beautiful extended family in Kolkata and Bangalore, comprising Krishna, Sudipta, Sumanta, Kaberi, Amita and Kabita Biswas, Pranab and Saswati Chakraborty, and Jyotirmoy Chakravorty. They supported me through the initial hurdles and kept me content through many happy and laughter-filled phone calls while I was in the UK writing up my dissertation through the uncertain COVID times. I take this moment to remember the blessings of near and dear family members, especially Sailesh and Gita Biswas, who did not live to see this book.

Finally, I would like to thank my family. My sister, Dishari, has been my one true critic; I remain ever grateful to her for painstakingly proofreading my work. Despite their busy schedule, she and Sayantan took the time to talk to me when I felt low. My in-laws, Lata and Bijon Dutta, constantly supported me. I cannot thank my parents, Santa and Parthasarathi Chakrabarti, enough for their endless encouragement and contribution to my academic career. They all let me stay at home and fed me while I was doing my fieldwork and constantly encouraged me to keep going. I could not finish my acknowledgement without mentioning the support of my partner, Saikat Dutta. He has been the most encouraging, pushing, yet motivating husband while I pursued my research and writing. Thanks, Saikat, for keeping me sane, feeding me and reading and listening to the numerous drafts of this book.

1 Durga Puja, Kumartuli, and Kolkata

Festival, religion, culture, and politics

In my fondest memories as a Bengali, born and raised in Kolkata, Durga Puja remains the most awaited event of the year. So, as a child, growing up closer to both my grandmothers, I grew up listening to their childhood memories of Durga Puja from villages in Bangladesh, cultural connotations, folklores associated with the festival, and a dominating perception of longing and nostalgia. To summarise, Bengalis, through generations, grow up with this deep-rooted association, often pride and awareness of religious worship that transcends to something bigger: a manifestation of the art, crafts, culture, and food as the festival of happiness and coming together. While there are other religious worships in the Hindu Bengali calendar around the year, none matches the scale of festivities around the annual Durga Puja in Kolkata.

Durga Puja is the worship of the Hindu female deity of Durga, during the autumnal months. It is the most popular festival in West Bengal and Kolkata's annual calendar. Durga deity is worshipped in different forms and times, such as in spring as *Durga* or as *Basanti* in different parts of the Indian subcontinent. However, the autumnal festival is the most widely celebrated among the Bengali populace, particularly in Kolkata and West Bengal. There are a few mythological and cultural accounts of why and how this autumnal religious festival started. According to Hindu mythology, Durga Puja is the victory of Goddess Durga over the demon of *Mahisasura*, and, according to the later Bengali version of the fifteenth-century Ramayana, Rama sought the blessing of Goddess Durga before his war by worshipping the deity (Simmons et al. 2018; Singh 2018).

In addition to the mythological accounts, my childhood memories included both my grandmothers narrating folklore, where Durga Puja is culturally associated with homecoming. The Durga deity, presented as the image of a daughter of the mountains in folk culture, visits her parental home from her marital house in *Kailash* (imagined somewhere in the higher snow-capped reaches of the Himalayas). According to elite upper-caste hegemonic discourses,[1] the Goddess Durga, also called *Mahisasur-mardini*, is represented to have killed the buffalo demon *Mahisasura* and triumphed over the evil powers to bring harmony and prosperity to the lives of the mortals on earth (McDermott 2001, Sen 2022). She is accompanied

DOI: 10.4324/9781003341222-1

Figure 1.1 Image of deities of Durga and her four children, *Lakshmi, Saraswati, Ganesh,* and *Kartik*

by her four children, *Lakshmi, Saraswati, Ganesh,* and *Kartik*, each a deity also widely worshipped in Hindu culture (Figure 1.1). Hence, Durga Puja generally brings about happiness derived from the gathering of the extended family and associated festivities, and thousands of Bengalis dispersed across the world customarily visit their hometowns.

In Kolkata, the religious deities are sculpted as idols of straw and clay, mostly in the historic neighbourhood of Kumartuli. Kumartuli, meaning potters' neighbourhood, is located along the Chitpur Road, a caste-homogeneous settlement that has existed for over two centuries. Every year, during the months leading up to Durga Puja, popular local newspapers publish dedicated articles on the goings-on in Kumartuli, especially how the preparations for the Durga idols are turning out. In such articles, journalists sometimes refer to the everyday struggles of the idol-making community, the crumbling infrastructure, the rising prices of items, happy memories of the olden days, and many more trivia. Such is the sentiment of Durga Puja among the local Bengali public that the Kumartuli neighbourhood is considered a brand that carries forward the tradition of idol-crafting and serving the city with idols for worship.

Durga Puja is a festival of mass local tourism interest. For decades, the festive installations locally known as *pandals* (marquee) housing hand-crafted clay idols of the deities and festive lighting attract millions of visitors to the carnival spread across Kolkata (Mukherji and Basu 2015). Heierstad (2017) opens his monograph by mentioning the scale of people's participation in Durga Puja festivities. 'Hundreds and thousands' of people from the suburbs and the nearby villages visit the city; it feels like 'no one is at home' during this time; public transport systems are overcrowded with curious visitors from across the world. The thousands of makeshift pavilions or *pandals* to house the clay idols, the temporarily widened footpaths (demarcated by bamboo fencing for crowd management), and traffic congestion due to overcrowded streets all add to the festivity of the city. The estimated worth of the economy generated through Durga Puja is US$5 billion, which includes not only the crafts economy related to idols and *pandals* but also festive paraphernalia, lighting, fashion retail, publishing and media communications, corporate advertisements, and much more (British Council 2019).

Durga Puja in Kolkata is paramount among other significant cultural and religious festivals. An estimated extra 200,000 to 300,000 people visit Kolkata per day alone during the event. There are about 28,000 community-sponsored pujas in the state (Dutta-Majumdar 2018), with nearly 2,000 in Kolkata (Chattopadhyay 2014), and more than 200 (Basu et al. 2013) such celebrations have budgets running into *crores* (ten million) of rupees, an estimate quite in contrast with the narrative of poverty that Kolkata holds. To illustrate the scale and participation in the communal pujas in Kolkata, Kolkata Police Force (KPF) reported that the metro railway services (within the city municipal area) carries approximately 3.3 million passengers, and the local railways carries 2.5 million passengers within and out of the city in the four days during the Puja (Basu et al. 2013). In fact, according to a 2019 report published by the West Bengal Tourism Department, the economy around this week-long festival accounts for 2.58 per cent of the state's GDP, which is about 32,377 crore INR (3.27 billion GBP) (British Council 2019). Also, thousands of seasonal informal migrant workers participate in the preparatory phase of this festival, working on idol-crafting and *pandal* construction.

However, it is fair to summarise that Kolkata's image is deeply entangled in the reflections of the religious and cultural renderings of Durga Puja festivities. The immense religious and place-based cultural connotations associated with the festival are reflected in the politics of the inscription of the festival in UNESCO's Intangible Cultural Heritage of Humanity representative list. Hence, not only has the religious worship been considered part of the inscription, but the place-based attribute that the city adds by being at the heart of the autumnal festivities is also inscribed as 'Kolkata's Durga Puja'. Sangeet Natak Akademi (SNA), India's apex body under the Ministry of Culture, nominated Kolkata's Durga Puja to be included in UNESCO's list (Sangeet Natak Akademi 2019). In its nomination of Durga Puja for the ICH Representative List, the SNA stated that:

> Durga Puja is the best instance of the public performance of religion and art in the city. It witnesses a celebration of craftsmanship, cross-cultural transactions and cross-community revelry. . . . The exemplary character of Durga Puja lies in its ability to not temporarily bind itself to the ritual occasion. Its dynamism lies in it being a constantly mutating event—in its fusion of

tradition with changing tastes and popular cultures, and in the adaptation of the iconographies of Durga and the styles of her temporary abodes to cater to new regimes of art production.

(Sangeet Natak Akademi 2019)

Under Chief Minister Mamata Banerjee, the current government has been unrelenting in its efforts to turn Durga Puja into a 'state event' (Guha-Thakurta 2017). Several events relating to the autumnal festival have been promoted directly by the state government, starting from maximising inaugurations by the Chief Minister to her painting of the eyes of the deity (ibid.). Also, there have been publicly funded projects on extensive documentation of Kolkata's contemporary Durga Pujas and the exhibition of photographs titled 'Bengal's Durga' in the Totally Thames Festival 2018 in collaboration with the British Council and the Tourism Department of West Bengal (British Council 2018). By far, the most prominent of these events is the immersion parade sponsored and organised by the West Bengal Tourism Department. Over 50 community-sponsored puja idols participate in the immersion procession during this rally. Several decorated lorries carrying idols and tableaux and neighbourhood groups adorning traditional clothes participate in the rally attended by dignitaries and celebrities, the regional media, and corporate houses (UNESCO 2019). However, the number of participating Puja committees seems insignificant as compared to the thousands worshipped in the city. While Durga Puja primarily remains the ritualistic worship of the deity, each year in Kolkata and adjoining areas, the scale of festivities and economic turnover seems to overtake the previous year's grandeur.

Colonial Calcutta's Durga Puja

Durga had been worshipped in Hindu families, limited only to the rich and upper castes before and during the early settlements in Calcutta in the colonial period. The rise and celebration of the Durga Puja in Kolkata are considered the reasons for the settlement of the *kumar* (potter) community in Kolkata starting from the mid-eighteenth century. These celebrations were a way to showcase the power and importance of merchants and officials (*babu* class) in the colonial era (Bhattacharya 2007). The pomp marked a status symbol of the gentry and the trading classes since the beginning of the settlement in Kolkata. Moreover, it is still a popular way of flaunting wealth and power; and organising a Durga Puja is even seen as a rite of passage in upper-class Bengali society. It is necessary to contextualise a historical account of how caste, class, and positions of power are interwoven into the dynamics of the growth and popularity of the festivities to unpack the beginnings of the ritualistic worship of the deity and the association of Durga Puja with the city of Calcutta.

Various records (Chaudhuri 1990a, Chaudhuri 1990b; Nag 1990; Bean 2011; Heierstad 2017) trace the origin of Kumartuli to the culture of celebrating the annual autumnal festival of Durga Puja. The festival has been historically celebrated in the eastern part of India, and initially, due to the elaborate rituals, it was only affordable to the wealthy, elite classes. Actively promoted and patronised by Maharaja Krishnachandra, Durga Puja was prominent in Krishnanagar (a few hundred kilometres north of the present-day Kolkata), which attracted a large number of devotees from the public to gather at the palace pavilion for ritualistic worship. In Calcutta, at the beginning of the eighteenth century, Durga Puja was a less popular event, confined only within the walls of *bonedi bari*s (mansions of the elites). One of the earliest socio-political commentaries of contemporary Bengal, *Hutom Pyanchar Naksha* (translated as Sketches by a Watching Owl), by Kaliprasanna Singha in 1861, narrated the emergence and evolution of Durga Puja in Kolkata and the growth of potters' settlement in Kumartuli (Nag 1990). This commentary has been integral to constructing the relationality of the case-based potters' settlement in Kumartuli to the ritualistic worship of Durga Puja among the elites in colonial Calcutta.

Shobhabazar Rajbari is a mansion in the *Shobhabazar* area close to Kumartuli. Nabakrishna Deb, the first and the most notable head of this family, was the '*Munshi*' (teacher of local language and trade) to Warren Hastings (the first Governor-General of India) (Deb 1990; Chattopadhyay 2005). Following the territorial conquest of then Bengal[2] and the win over Nawab Siraj ud-Daula (Chaudhuri 1990a) in 1757, Robert Clive (chief military commander of the British East India Company) agreed to celebrate his victory at Nabakrishna Deb's mansion (Chaliha and Gupta 1990). This puja was key for Nabakrishna to maintain his social and political position and continue his business. The celebration of Durga Puja by Nabakrishna at the *Shobhabazar Rajbari* subsequently became the 'company puja' for the natives (Mitra 2015). In order to manifest their powers and a position of dominance in the local elite societal realms, the annual *Shobhabazar Rajbari* puja would first start their ritual in Kolkata, marked by cannon fire, and the rest of the city could only start afterwards (Chaudhuri 1990a). It is essential to highlight how the social and political positions were asserted and grew around the Durga Puja celebrations from the very beginning. These earliest customs remain significant and influenced the elite and subsequently the average Bengali to shape socio-religio-political identities to this day.

Similarly, the emerging *babu*[3] class, who were mostly involved in trade with the Europeans in Kolkata, celebrated their family ritual of Durga Puja with much grandeur (Deb 1990; Mitra 2015). The festivities at their residences would aim to flaunt their abundant wealth and be a means to liaise with their then British clientele. In addition to the traditional household rituals for the festival, cultural differences among the British and the native Bengalis often resulted in these *babu* households having separate celebrations intentionally aimed to impress their seemingly superior British invitees. Contradictory to the traditional customs of a Hindu Bengali household ritual, often, the celebrations were attempted solely for the entertainment of the British officers; these late evening concerts included song and dance by the *baiji*,[4] *jatrapala*[5] or play, and served meat and wine (Bhattacharya 2007). Women of the household were possibly not permitted to attend these functions. There seemed to be gendered roles in these celebrations; men mostly attended to the entertainment of their guests while the women oversaw organising the traditional ritualistic aspects of the festival with the help of the servants.

The women of the household were instead involved in preparing for the Durga Puja. They dyed and made the *sarees* and other festive paraphernalia for the idols and households themselves (Chaliha and Gupta 1990). Idol-makers in those days used to prepare the idols in the pavilion or courtyard within the family mansions. They were accompanied by *malakars* (persons belonging to the florist and garland-weaving caste). This custom of familial involvement in the practice of idol-making continues till today; interestingly, the same *kumar*-family through generations serves the same *babu* household with a similar style of the idol (interview K9). Although the *kumars* now must craft idols for other clients for their livelihoods, they follow the custom of serving the same *babu* family to maintain their family reputation and existing clientele. The framework or the *kathamo* for the idol is worshipped first on the day of *Rathayatra* (a festival in monsoon, usually in June/July), following which the idol-making process (described in detail in Chapter 2) begins for the Autumnal celebration of Durga Puja (late September/October). Notable Bengali ritual calendar events related to idolatry are tabulated in Table 1.1 in this chapter. *Kumars* from modern-day Kumartuli still visit their '*babus*' in North Kolkata, and it is from these *babu*-families' patronage that some *kumars* earned fame among the British and subsequently the wider public. Such traditions have survived through generations and oral histories, much of which has been reflected in my interviews with the *kumars*.

Table 1.1 The annual pattern of idol-making in the Kumartuli area

Gregorian Months	Bengali Months	Seasons	Festivals	Approximate Preparation Phase and Additional Labour Requirement
December–February	*Poush–Magh*	Winter–spring	Sarasvati puja, Gopal puja	1 month
March–May	*Phalguna–Chaitra–Boishakh*	Spring	Basanti/Annapurna puja, Ganesh puja	1 month
May–July	*Boishakh–Jaishthya–Ashar*	Summer	Rathayatra	0 month
July–September	*Ashar–Shrabon–Bhadra*	Monsoon	Viswakarma puja, Manasa puja	2 months
September–November	*Bhadra–Ashwin–Kartik*	Autumn	Durga Puja, Lakshmi puja, Kali puja,	>6–8 months
November–January	*Kartik–Agrahayan–Poush*	Late autumn–winter	Kartik puja, Jagaddhatri puja	3–4 months

Source: tabulated using data from the Bengali calendar and interviews at Kumartuli

Originally part of one notable *babu* household of Kolkata, the first woman entrepreneur, Rani Rasmani, changed the conventional practices and philanthropic patronage radically during Durga Puja (Chaliha and Gupta 1990; Chaudhuri 1990a). She endorsed more involvement with the local community in her family's Durga Puja celebration within their mansion and distributed gifts to the underprivileged. She supported local talent in her all-night *jatras* and other cultural programmes instead of only entertaining the British dignitaries in the then Calcutta. These cultural functions were perhaps instrumental for later wider social involvement of the Bengali public in the Durga Puja celebrations (Chaudhuri 1990b).

Barowari brought inclusivity

Until the mid-eighteenth century, despite public interest, the ritualistic aspects of the Durga Puja were primarily confined to the domestic circles of the wealthy and elite upper classes. Some neighbourhoods and groups of merchants felt the urge to arrange public facing festivals (possibly through cooperative funding) due to the increasing need for community pujas outside the exclusive *babu*-houses. Although there is some uncertainty in the historical accuracy, the general consensus is that 12 (*'baro'*) friends (*'yar'*) near Guptipara (in the Hooghly district of Bengal) in around the late eighteenth century facilitated the first community arrangement of performing the worship through communal subscription. However, this subscription-based celebration too was a highly upper-caste event (Mitra 2015). Although the records are not relatively abundant and somewhat inconclusive, the date of the first *baroyari* puja is between 1761 and 1790. This is an important event as it paved the way for public participation in the future festivities of Durga Puja. Subsequently, all community-sponsored subscription-based participatory pujas have been emphatically called *baroyari* or *barowari* (pronunciation based on dialectic variations) puja.

Towards the end of the eighteenth and early nineteenth century, community-sponsored worship of Durga Puja became reasonably popular. The *barowari* puja gained popularity among the masses and was open to all, somewhat democratising the participation of the common person and his humble family. With the rise of *barowari* pujas, the commissioned *kumars* were restricted to preparing the idols in their workshop within their family home, unlike the previous confines of the then *babu* mansions (Ray 2017). *Kumars* consequentially had to make room for preparing the idols within the already limited space of their home-based workshops. Eventually, the limited means of the *kumars* and the cramped conditions of their workshops in the Kumartuli settlement started becoming cultural production spaces, gradually securing international acclaim.

Temporary pavilions or marquees (*pandals*) began to be erected for the festival days to protect the unfired clay idols from nature during the days of the festival. The *pandals* serve the purpose of housing idols as well as a congregation space for the community during worship. Around 1911, *barowari* puja was renamed or rebranded as '*Sarbojonin*' (for everyone, all-inclusive, and for all to participate) puja by the involvement and patronage of the then Indian National Congress

(Sarma 1969; Bhattacharya 2007). The rebranding implied that it was planned to be an inclusive attempt to appeal to a wider section of society irrespective of caste and class. The Indian National Congress started using the public Durga Puja as part of the nationalist movement and demonstrated a solidarity pledge for the freedom movement in India. Festivities around the *pandals* promoted the sale of native goods and employed significant numbers of local merchants. Shifting from the pompous domesticity of *babu* mansions, evening functions in *Sarbojonin* pujas involved local youth participating in nationalist songs and plays. Participation in the nationalist movement promoted through the Durga Pujas gained popularity and drew the wider public. The first *Sarbojonin* Durga Puja organised in the Kumartuli neighbourhood was in 1931. This celebration was backed by the then Indian National Congress President Subhash Chandra Bose, which attracted mass participation (Chaudhuri 1990b). Hence, within a century, the Durga Puja, from being a domestic ritualistic worship restricted to the wealthy, became a widely participated festival that united masses to mobilise against a common threat: the colonial government.

Durga Puja evolved as the city of Kolkata grew. Starting from the few celebrations among the elites in the early eighteenth century, Durga Puja slowly spread to middle-class neighbourhoods and became *Sarbojonin* (inclusive). The demand for idols for Durga Puja and for household worship of other festivals urged potters to make idols eventually. Subsequently, the *babu* mansions concentrated in and around the northern quarters of Kolkata were not the only clientele of the *kumars* of Kumartuli. The once-influential *babu*-families of Kolkata might not have been able to retain their businesses and fame, but they were instrumental in setting a lasting image of festivity and culture in public memory and perception of Kolkata. Historical documents and paintings depict the grandeur of these festivities (Chaudhuri 1990a; Bhattacharya 2007; Ray 2017). The *babu*-family idols, which a few *kumars* of Kumartuli still have to prepare in the courtyards of the old mansions of Kolkata, now add promotional value to their businesses. Moreover, the lasting image of the *babu*-households and older influential community-sponsored pujas are a means of advertisement and branding of Kumartuli artists. Most *kumars* work with their 'old customers' on a 'generational basis'. In an interview with one of the older *barowari* Puja committees in this area, the *Bagbazar Sarbojonin* Durgotsav [the literal translation which the organising committee members stressed upon means participatory Durga festival] suggested that they maintain their standard of celebration as it used to be done in the past. This Puja committee, which marked its 150th anniversary in 2018, has maintained the purchase of their idol from the same line of *kumar* family.

Due to India's partition and after independence, many Bangladeshi *kumar* families arrived in Kumartuli. The migration of Bangladeshi potters has helped the community to grow and mingle two slightly different styles of idol-making. The Bangladeshi potters' style of idol-making, commonly called the *Bangla* style, varied mostly from the already existing *Sabeki* (traditional) Kumartuli style (Figure 1.2). Noticeable differences were the features of the face and the figurative proportions of the female deities; these were fuller in shape and looked rather anthropomorphic than the ones previously sculpted in Kumartuli. The idols' colour

Figure 1.2 Images of idols: (a) traditional (*Sabeki*) style, the *Bagbazar* idol, and (b) more human-like (Bangla) style of artists of Bangladeshi heritage

and the eyes' style were more anthropomorphic than the artistic renditions of the previous Kumartuli style. Many idol-makers, irrespective of their origin, have successfully earned fame under the branding of their own names; the names carry their ancestry and area of origin. Their names carry their caste and relate to their ancestry in popular realms and their area of origin. That adds to their identity. Heierstad (2017) argues that the *maliks* (the now-famous *kumar*, owner of a business) affiliated to the culturally embedded profession are part of the tradition endowed by their caste. However, their '*imaginations have been freed*', and they are involved in creating outstanding art images celebrated by the Bengali culture. The clients with higher budgets of celebration, like *Bagbazar*, know that sculpting the idol of *Bagbazar* is important for a *kumar*'s 'profile' or portfolio. The family associated with this idol-making have maintained their unique facial features and structure style and are obliged to carry on the same style for generations on the client's demand. Although this family has gathered other notable clients, they continue to be commissioned for the same style of an idol because of their fame for the same Sabeki Bagbazar-idol style. The organisers believe that the public image of the idol has forced the *kumar* to follow the traditional style for this idol, which is his 'USP' (what they loosely translate to a unique selling point) or what sets them apart from the rest, they added.

Cultural heritage, informality, and idol-crafting practice

This book is about how this historic inner-city, situated, religious idol-crafting community in India is undergoing continuous socio-spatial transformation due to an access to physical and social infrastructure, local governance policies, socio-political hierarchies, and complexities of informal tenure. Kumartuli, from its inception, has been an informal settlement often known in Urban Studies and related literature as a slum.[6] Cultural heritage rooted in sociological studies and broadly defined by UNESCO as the valuable, customary, and traditional artefacts, buildings, and monuments carrying social and anthropological significance outlines the framing of this book. While different connotations of the cultural policies, festivities, and their implications are central to this analytical framing, it is also imperative to consider these in relation to the largely informal nature of the cultural industries and the slum neighbourhood embedded within this network. Hence, implications of such religio-socio-political rhetoric form the relational place-based understanding of a heterogeneous urban space within a tenured slum settlement.

Informal settlements in inner-city areas in the global South continue to be spaces of contestation in Urban Studies literature (Watson 2009; Simone 2019). While many debates grapple with the marginalisation of informal communities and their rights to the city and infrastructure (Anand 2012; Truelove and Cornea 2020), academic literature also presents the 'creativity' within these largely underrepresented sections of the city (Badami 2018; Mbaye and Dinardi 2019). It is often found that inner-city areas are densely inhabited spaces of social and material practices often marginalised from mainstream governance due to complexities of informality and lack of representation. This book is about the inner-city neighbourhood of Kumartuli, where religious idols for Hindu festivals are crafted, and presents the politics of marginalisation and the changing of consumption practices. This research, based on an informal setting in Kolkata, builds on the need to empower marginalised community residents to contribute to decision-making processes within policy realms. The theoretical, methodological, and empirical components of the research contribute to ongoing debates on citizenship, informal settlements, and the role of cultural industries-led regeneration policies in inner areas of Southern cities.

The book investigates how people interact and adapt to existing social and physical infrastructures. Their interaction is more complex than what is often expressed in literature. Focusing on a neighbourhood where everyday lives are intertwined with larger informal enterprises, this book discusses how the logistics of the seasonal economic productive function are accommodated within the everyday practices of residents. This conceptual framing of this book elaborates on the theoretical underpinnings of social and spatial practices concerning the Kumartuli neighbourhood and the relevance of the ongoing cultural policies, such as the UNESCO's ICH inscription and wider promotion of Durga Puja as part of the local political agenda. Hence, this book grapples with different yet complementary strands of concepts.

Theories of practice (Shove et al. 2012) provide a platform to incorporate challenges faced in everyday life and in a way to amalgamate the contestation, coordination, and competitions within and between coexisting practices. Spaces are defined and redefined by practices and communities of practice where boundaries are interwoven and often are blur (Schatzki 1996). This challenges our understanding of practices coexisting in constrained production spaces such as slums. Subsequently, the spaces might be restructured to fit the changing needs and demands of such practices based in dense urban areas of the global South. Although the dynamic of social practices framework (Shove 2016) addresses the idea of individuals as carriers of practice who adopt and adapt to the locally changing social meanings of practices, social inequalities, geographical constraints, or cultural diversity of the empirical examples of such practices are not mentioned. In a way, the empirics do not cross the geographical and historical boundaries of the recent past of the developed West. However, we might be able to theorise practices from a Southern theory perspective (Bhan et al. 2018; Lawhon 2020) by studying the diverse elements of practices and the materiality of spaces and places relationally.[7]

In the case of studying the idol-making practices based in Kumartuli, it must be highlighted that the meaning and significance of the idol are changing based on the technical know-how of idol-making required for branding and selling to a wider consumer base. This study focuses on the evolving practices that have shaped the working conditions of the idol-makers' community in Kumartuli and their corresponding living conditions in this place for generations. The community's everyday social and cultural practices overlap with the idol-making practices within similar spatial and temporal frames. Practice theory, in a way, acknowledges that practices are place-based, but the aspect of space remains somewhat ambivalent. Using the practice theory framework to understand spaces and places away from the widely applied and discussed scenarios of the West calls for an amalgamation of different sets of literature, such as the Southern cities' perspectives (Robinson and Roy 2016; Lawhon 2020) and place-based literature (Lombard 2014). Also, theoretical and cultural assumptions of the West do not fit in with the cases of postcolonial cities (Roy 2009; Robinson 2016; Lawhon and Le Roux 2019). This book presents a better understanding of the practices of *basti* (or *bustee*)[8] (Bhan 2017) communities in older Indian mega-cities with an attached economic activity and their socio-spatial characteristics that maximise the place-identity. This research challenges the dominant perception of theorising informality. In doing so, it contends undifferentiated views of informalities from the South (Banks et al. 2020). I question whether there is a scope for understanding the informal sectors through more nuanced unbiased framings. It does not accept the derogatory notion of slums and squatters, nor does it try to establish creativity in slums, but it tries to investigate the social, cultural, and spatial practices relating to the wider political economy, which are unique in *bastis* of Southern cities. Following on from Bhan (2019), I argue that there is a need to look at informality relationally from the Southern perspective with regard to the social, cultural, and spatial practices.

The book unfolds the complexities of caste, gender, class, and political identities and affiliations associated with the multiple factors of inclusion and exclusion (Datta and Ahmed 2020), particularly in the case of access to infrastructure in informal settlements (Desai et al. 2020) in urban areas with an added productive function. Also, this research debates how cultural tourism policies and subsequent spatial transformations might threaten the existing practices in an inner-city slum. Through an innovative place-based analytical lens, it questions policies of regularising or redeveloping informal settlements in tackling the housing crises and existing infrastructure in such areas of thriving informal sector economies.

Besides the issues of marginalisation due to the location and character of the neighbourhood, certain cultural policies and practices continue to be somewhat exclusionary. For example, the most recent of such policies are the major initiatives of the local government of West Bengal regarding the promotion of Kolkata's Durga Puja and inscription in the ICH list. While focusing on the idol-making community, these efforts also highlight the larger stakeholders of the economy surrounding the festivities. The nomination document highlights the involvement of several stakeholders, including international organisations, corporate houses, and national media (UNESCO 2019). Additionally, they have partnered with international organisations, namely the British Council, to promote tourism (British Council 2018). The nomination of Durga Puja in the ICH list initiative not only raises questions as to how the idol-making practices based in Kumartuli are incorporated within the ICH framework but also draws on important debates of urban development and inner-city spatial policies that threaten marginalised communities. Contrarily, regional policies such as West Bengal's creative and cultural industries promotion put forth an agenda to enhance entrepreneurship within this sector. The Government of West Bengal's Tourism Department has also proposed policies promoting tourism through the creative cultural industries and crafts, like weaving, textiles, and clay idols, which are traditional to Bengal, where international collaborations are in place to inform policy. Such efforts, seemingly informed by western urban regeneration models through culture, creativity, and creative economy,[9] do very little to address inequalities in Southern cities. I must clarify that while Kumartuli is at the centre of the relevant policies that promote traditional cultural industries, these policies might threaten the practices and livelihoods of residents. Despite Kumartuli's branding, the representation of the potters' community is quite low, and there remains a narrative of marginalisation among the potters' community. Additionally, the involvement of UNESCO and using tools like culture and heritage to ascertain sustainability essentially put an obligation on these international actors to learn and invest in place-based knowledge.

Qualitative techniques form the basis of this research to critically engage with generating and analysing empirical data gathered through detailed ethnographic study. This research is situated at the intersection of architectural and human geography research. Rooted in questions on relationships between everyday social practices, spatiality, and the material processes in place to foster these relations, architectural and human geography research tools and techniques were combined to achieve the research objectives. Human geography research methods influenced the exploration of the place (Flowerdew and Martin 2005). Architectural research methods with a qualitative approach to suit the case study were at the centre of a conceptual framework (Groat and Wang 2013).

This research design is presented as a methodological appendix and illustrates the use of a combination of tools and techniques that were developed through my interest in understanding the flows and processes of idol-making practices. The theoretical and methodological frameworks of the research were developed through iterations and feedback from challenges faced during fieldwork. Upon finalising the case study and theoretical framework, I adapted the aims, objectives, and respective methods through various iterations. Designed around a detailed case study, a series of qualitative techniques ranging from ethnography-inspired approaches of semi-structured interviews, visual documentation, mapping, to a participatory visual method is broadly at the core of this research design. The research entails a detailed study of practitioners, stakeholders, and institutions involved in idol-making and the subsequent field mapping, through interviews, document analysis, and discourse analysis. During my doctoral study, I carried out fieldwork in two phases for seasonal and logistical purposes, such as witnessing the peak production and festive phase, while also finding intermediate phases for interaction with research participants. Finally, I again went back to Kumartuli during my postdoctoral fellowship to conduct a deliberative workshop solely to follow up on infrastructure and public services-related discontents of my research participants in Kumartuli.

Participant observation study was key to documenting the process of production. I started with mapping the wider relational geographies associated with idol-making through the multi-sited ethnography-inspired approach of 'follow the idol'[10] and documenting the processes and flows of the industry based primarily in Kumartuli. This process introduced me to the entire network and helped me map it out for future analysis. Participants in the idol-making community-based in two adjacent locations, because of the stalled planned intervention in Kumartuli, were crucial to identify the network of supply chain and the actors involved to document the same. I also visited and documented a few sources of the basic raw material and labour around southern Bengal. However, documenting processes over a wider area and interviewing actors proved difficult. While most interviewees were ready to talk about general information on the idol-crafting process and traditional practices, they were hesitant to talk about current issues and challenges they faced since the failed redevelopment project. The second phase of the fieldwork was primarily informed by the first phase and designed mostly around its feedback. During phase two of my fieldwork, I concentrated more on the neighbourhood. I studied in detail the spaces of production, the building typologies, and the different spatial reconfigurations happening because of the increasing demand and growth of the idol-making industry. My biggest challenge during both phases was recording interviews, particularly on questions about buildings, workshop spaces, tenures, and infrastructural challenges. Interviewees seemed hesitant to speak about their everyday challenges, fearing being overheard by neighbours or passers-by, a fear stemming from the local politics and power hierarchies. Hence, I had to improvise my method to use a more participatory visual method to overcome this challenge. Participatory visual studies (Davis et al. 2018) provided necessary interactions with the community while building on its cultural and social assets. Using participatory photography and photo elicitation techniques, also known as PhotoVoice[11] within the community, was a means to gather data on place-based notions of the participants. Participants took photographs of their surroundings and their working and living environments to evoke meanings of the place and how they identify with it.

Structure of the book

The central focus of this monograph is investigating the material and spatial and social changes in Kumartuli. Using social theories of practice (Shove et al. 2012; Shove 2017), I explore cultural practices and spaces, both domestic and public, and study how idol-making is shaped by and has shaped the built form and infrastructure of the neighbourhood. Through the different chapters, this book presents how the idol-crafting community in Kumartuli has adapted to the existing infrastructure to facilitate the idol production and distribution with accommodating the seasonal surges in worker numbers, and how, eventually, the mixed-use residential neighbourhood with its established network is changing to a more commercial one. This transition is being facilitated in part due to the informal nature of the land tenure and notified slum status of the neighbourhood (Kundu 2003; Bhan 2019; Banks et al. 2020).

The monograph questions if Kolkata has succeeded in acknowledging and catering to the needs of idol-makers and other marginalised creative communities embedded in inner-city slums. This work establishes the need for incorporating the place-based practices of such traditional crafts industries in the cultural policy domains. Combining these findings and questions, it completes a further round of investigations on how residents of a densely populated slum interact with the existing uneven physical infrastructural systems and overcome these precarities. So far, starting from the image of Kolkata, associated with Durga Puja and the politics of nominating and subsequent listing of the festival for UNESCO's ICH, this chapter has drawn the conceptual frameworks used in this research to investigate the practices of emerging allied new communities of idol-makers and seasonal workers, which cater to the Durga Puja festivities.

Chapter 2 of this book introduces the traditional idol-crafting practice, how it benefits from the geographic and urban location, where materials came from historically, and how these are procured now. It first details the age-old sustainable materiality of idol-crafting and worshipping practice and how that relates to spiritual values and norms. Then it begins to uncover the reasons for the contemporary deviations in the crafting practice due to Durga Puja's changing consumption patterns, spectacular and competitive festive installations, and tourism pressures. It details the scale of production, transactions, and exchanges while presenting the wider political economy of Durga Puja festivities. This chapter discusses the unsustainable consumption and practices slowly displacing traditional practices to give way to newer iconic idol-crafting that satisfies the tourism and government's vision. This chapter builds on ethnographic evidence from fieldwork and existing anthropological literature on the evolution of the craft and discourse on cultural and religious politics.

Chapter 3 presents the built environment and the socio-material infrastructure that enables and/or hinders the idol-crafting practice. Drawing on ethnographic and architectural studies in Kumartuli, this chapter presents a detailed analysis involving drawings, maps, spatial layouts, and photographic materials of workspaces, streets, riverfront embankments, and buildings. The descriptions also comment on how idol-making practices have shaped, adapted, and evolved around these spaces. This chapter details the spatial layout of individual workshop-residences typical to Kumartuli and discusses how traditional practices and age-old norms of gender, class, and caste govern these cramped yet flexible spaces within an informal setting. This chapter specifically demonstrates how material religious and spatial practices in this idol-crafting neighbourhood have a unique socio-spatial positioning in the urban fabric of Southern cities.

During the idol-crafting season before the festival, public spaces and infrastructural shortcomings are adapted, adopted, or repurposed by the craft's practitioners to carry out the practice. Chapter 4 draws on the dynamic between facilities being stretched to breaking point and people's grievances and the continuing faith-led consumer demand for clay crafts. That is, a scholarly understanding of Durga Puja as a religious practice and Kumartuli as a crafting neighbourhood must be located against the complex backdrop of the growing commodification of a cultural craft, as well as an understanding of how the seasonal workforce accompanied by festive drummers, and associated caste-based professions and other networks have evolved to facilitate these operations of practices. Findings suggest that despite the congested and competitive spatial and relational configurations, social cohesion, collaborative practices, and support networks within the caste-homogeneous neighbourhood sustain the growing demand for crafts associated with Durga Puja.

Chapter 5 details how a failed redevelopment project by the local government in 2009 left the neighbourhood in an infrastructural limbo. It develops around the details of the land and building ownership complexities in an inner-city slum and how rights and tenures are impacted by multiple exclusionary factors such as location, age of settlements, caste, class, and gender. This monograph stands out particularly from the rest of the scholarly works on Kumartuli and idol-crafting, mentioned later, in its unpacking of the built character, land and building complexities and infrastructural failings, as well as how despite these multiple marginalising aspects, the community continues to get by and get on. These discussions draw on wider debates on Southern urban practices (Bhan 2019), infrastructural scholarship in recent decades (Amin 2014; Cass et al. 2018), and aspects of inclusion and exclusion around these (Chattopadhyay 2012; Datta and Ahmed 2020).

Chapter 6 details the trends in the changing spatial character of Kumartuli's idol-crafting spaces and how traditional workshop spaces are slowly being replaced by commercial spaces detached from residences while the families are displaced. This chapter unfolds the emerging types of commercial workshops, repurposed to accommodate the growing production numbers, picking up from the discussion in Chapter 4. This chapter assesses the architectural detailing of production spaces and the associated practices performed within these spaces.

To conclude, this book questions whether the local and central government policies, including UNESCO's ICH nomination and the subsequent listing, have been able to acknowledge the needs of the idol-making and allied communities within this largely informal settlement of Kolkata's historic inner core. In doing so, Chapter 7 draws on debates on the future of crafts/cultural industries embedded in the Southern city's fabric and the need to empower these communities and incorporate them within the policy realms. These debates build on ongoing discussions concerning citizenship, marginalisation, land and building ownership among the urban poor, and rights to infrastructure emerging from the South.

Notes

1 While Durga Puja is symbolic in religious and political discourses of contemporary India, particularly in Bengal, many caste minority groups in India worship *Mahisasura*

as a historical figure which is a significant subversive narrative. See Sen, M. (2022). From demon to deity: Forging a new iconography for Mahishasur. *Journal of Material Culture*: 13591835221116708.

2 Territorial limits of the then Bengal included parts of present-day West Bengal, Bangladesh, Bihar, and Odisha and was the first military victory of the British East India Company in India, hence an important event in the history of British colonial rule in India.

3 Babu is the aristocratic, well-dressed man of the emerging Bengali wealthy merchant class, whose fortunes were built by trading with the English East India Company and other overseas trades.

4 *Baiji* is a professional woman singer and dancer trained in Hindustani classical music.

5 *Jatrapala* or *jatra* are Bengali equivalent of opera.

6 The word 'slum' could be crudely defined as a compact settlement with a collection of poorly built tenements, mostly of temporary nature, crowded together usually with inadequate basic services and subject to unhygienic conditions. Various agencies, including international organizations like UNHABITAT, have defined 'slum' in different ways, depending on the purpose and issues under consideration. However, there are certain broad similarities in definitions adopted by countries across the world.

7 The Southern city critique leads to the ongoing debates in human geography and urban planning about the relational aspects of place and the need for understanding places as dynamic and open. For more discussion on the Southern urbanism critique on specificities and generalisability, see Mohan, A.K. (2021). Introduction—Exploring urban 'southernness': Praxes and theory(s). In Mohan, A.K., Pellissery, S., & Aristizábal, J.G. (Eds.), *Theorising Urban Development from the Global South*. Cham, Switzerland, Palgrave Macmillan. 1–28.

8 A colloquial term, used widely in India, and hence acceptable contradistinctive alternative to the word slum. While 'slums' have certain negative connotations associated with the usage, *basti* being a widely referenced term within the society is more acceptable.

9 Since 1990s, creative economy has been a tool to regenerate post-industrial urban landscapes, particularly in the UK. See Warren, S. and P. Jones (2016). *Creative Economies, Creative Communities: Rethinking Place, Policy and Practice*. London and New York, Routledge.

10 The methodological approach in ethnographic research, which involves the tracing of objects, products, or people located around multiple geographical locations transnationally, forms the basis of multi-sited ethnography. The supply chain of a product mapped through global consumerism is an emerging trend in interdisciplinary studies including media, human geography, science, and technology studies. This type of ethnographic study involves research carried out in multiple locations and addresses geographies of production, transportation and consumption of commodities, and the underlying narratives of exploitation of the labour market. See Marcus, G.E. (1995). Ethnography in/of the world system: The emergence of multi-sited ethnography. *Annual Review of Anthropology*. 24: 95–117, and Knowles, C. (2014). *Flip-Flop: A Journey Through Globalisation's Backroads*. London, Pluto Press: 6–8.

11 See Wang, C. and M.A. Burris (1997). Photovoice: Concept, methodology, and use for participatory needs assessment. *Health Education & Behavior*. 24(3): 369–387. And, Sutton-Brown, C.A. (2014). Photovoice: A methodological guide. *Photography and Culture*. 7(2): 169–185.

References

Amin, A. (2014). Lively infrastructure. *Theory, Culture & Society*. 31(7–8): 137–161.

Anand, N. (2012). Municipal disconnect: On abject water and its urban infrastructures. *Ethnography*. 13(4): 487–509.

Badami, N. (2018). Informality as fix: Repurposing Jugaad in the post-crisis economy. *Third Text*, 2–4 Park Square, Milton Park, Abingdon OX14 4RN, Oxon, England, Routledge Journals, Taylor & Francis Ltd. 32: 46–54.

Banks, N., D. Mitlin and M. Lombard (2020). Urban informality as a site of critical analysis. *The Journal of Development Studies*. **56**: 223–238.

Basu, S., I. Bose and S. Ghosh (2013). Lessons in risk management, resource allocation, operations planning, and stakeholder engagement: The case of the Kolkata Police Force and Durga Puja. *Decision*. **40**: 249–266.

Bean, S.S. (2011). The unfired clay sculpture of Bengal in the artscape of modern South Asia. *A Companion to Asian Art and Architecture*. Malden, MA, Wiley Online Library: 604–628.

Bhan, G. (2017). From the *basti* to the 'house': Socio-spatial readings of housing policy in India. *Current Sociology*. **65**: 587–602.

Bhan, G. (2019). Notes on a Southern urban practice. *Environment and Urbanisation*. **31**: 639–654.

Bhan, G., S. Srinivas and V. Watson (2018). Introduction—The Routledge companion to planning in the Global South. *The Routledge Companion to Planning in the Global South*. London, Routledge.

Bhattacharya, T. (2007). Tracking the goddess: Religion, community, and identity in the Durga Puja ceremonies of nineteenth-century Calcutta. *Journal of Asian Studies*, Cambridge University Press. **66**: 919–962.

British Council (2018). Bengal's Durga at Totally Thames 2018. *British Council*.

British Council (2019). Mapping the creative economy around Durga Puja 2019. British Council.

Cass, N., T. Schwanen and E. Shove (2018). Infrastructures, intersections and societal transformations. *Technological Forecasting and Social Change*. **137**: 160–167.

Chaliha, J. and B. Gupta (1990). Durga Puja in Calcutta. *Calcutta The Living city: Vol II*. S. Chaudhuri. Oxford, Oxford University Press. **2**: 331–336.

Chattopadhyay, S. (2005). *Representing Calcutta: Modernity, Nationalism and the Colonial Uncanny*. Oxford, Routledge. **2**.

Chattopadhyay, S. (2012). *Unlearning the City: Infrastructure in a New Optical Field*. Minnesota, University of Minnesota Press.

Chattopadhyay, S. (2014). Politics, planning, and subjection: Anticolonial nationalism and public space in colonial Calcutta. In Chattopadhyay, S., & White, J. (Eds.), *City Halls and Civic Materialism*. London and New York, Routledge: 221–238.

Chaudhuri, S. (1990a). *Calcutta, the Living City: The Past: Vol I*. Oxford, Oxford University Press.

Chaudhuri, S. (1990b). *Calcutta, the Living City: The Present and Future: Vol II*. Oxford, Oxford University Press.

Datta, A. and N. Ahmed (2020). Intimate infrastructures: The rubrics of gendered safety and urban violence in Kerala, India. *Geoforum*. **110**: 67–76.

Davis, A., A. Javernick-Will and S.M. Cook (2018). A comparison of interviews, focus groups, and photovoice to identify sanitation priorities and increase success of community-based sanitation systems. *Environmental Science: Water Research & Technology*. **4**(10): 1451–1463.

Deb, C. (1990). The 'great houses' of old Calcutta. *Calcutta The Living City: Vol I*. S. Chaudhuri. Oxford, Oxford University Press. **1**: 56–63.

Desai, R., D. Mahadevia and S. Sanghvi (2020). Urban planning, exclusion and negotiation in an informal subdivision: The case of Bombay Hotel, Ahmedabad. *International Development Planning Review*, Liverpool University Press. **42**: 33–56.

Dutta-Majumdar, A. (2018). *Mamata Banerjee Announces ₹28 Crore 'Gift' for Durga Puja Organisers*. Livemint.com. Retrieved June 6, 2023, from https://www.livemint.com/Politics/0ZYs3G8oDzx0vBPgEBF7RM/Mamata-Banerjeeannounces-Rs-28-crore-gift-for-Durga-puja.html.

Flowerdew, R. and D. Martin (Eds.) (2005). *Methods in Human Geography: A Guide for Students Doing a Research Project*. London, Pearson Education.

Groat, L.N. and D. Wang (2013). *Architectural Research Methods*. Hoboken, NJ, John Wiley & Sons.

Guha-Thakurta, T. (2017). *In the Name of the Goddess: The Durga Pujas of Contemporary Kolkata*. Retrieved June 6, 2023, from https://www.thehindu.com/profile/author/Tapati-Guha-Thakurta-12490/.

Heierstad, G. (2017). *Caste, Entrepreneurship and the Illusions of Tradition: Branding the Potters of Kolkata*. London, Anthem Press.

Knowles, C. (2014). *Flip-Flop: A Journey Through Globalisation's Backroads*. London, Pluto Press: 6–8.

Kundu, N. (2003). The case of Kolkata, India. *Understanding Slums: Case Studies for the Global Report on Human Settlements*. Nairobi, United Nations.

Lawhon, M. (2020). *Making Urban Theory: Learning and Unlearning Through Southern Cities*. London and New York, Routledge.

Lawhon, M. and L. Le Roux (2019). Southern urbanism or a world of cities? Modes of enacting a more global urban geography in textbooks, teaching and research. *Urban Geography*. **40**(9): 1251–1269.

Lombard, M. (2014). Constructing ordinary places: Place-making in urban informal settlements in Mexico. *Progress in Planning*, Elsevier. **94**: 1–53.

Marcus, G.E. (1995). Ethnography in/of the world system: The emergence of multi-sited ethnography. *Annual Review of Anthropology*. **24**: 95–117.

Mbaye, J. and C. Dinardi (2019). Ins and outs of the cultural polis: Informality, culture and governance in the global South. *Urban Studies*. **56**: 578–593.

McDermott, R.F. (2001). *Mother of My Heart, Daughter of My Dreams: Kālī and Umā in the Devotional Poetry of Bengal*. New York, Oxford University Press on Demand.

Mitra, A. (2015). *Kolkata o Durga Pujo*. Kolkata, Ananda Publishers Private Limited.

Mohan, A.K. (2021). Introduction—Exploring urban 'southernness': Praxes and theory(s). In Mohan, A.K., Pellissery, S., & Aristizábal, J.G. (Eds.), *Theorising Urban Development from the Global South*: 1–28. Cham, Switzerland, Palgrave Macmillan.

Mukherji, A. and S. Basu (2015). Crisis in representationd and reading: 'It's all Rheydt', Kolkata, 2011. *Journal of South Asian Studies*. **03**(02): 261–273.

Nag, A. (1990). *Sateek Hootum Pyanchar Naksha (Bengali Edition)*. Kolkata, Subarnarekha.

Ray, M. (2017). Goddess in the city: Durga pujas of contemporary Kolkata. *Modern Asian Studies*. **51**: 1126–1164.

Robinson, J. (2016). Thinking cities through elsewhere: Comparative tactics for a more global urban studies. *Progress in Human Geography*. **40**.

Robinson, J. and A. Roy (2016). Debate on global urbanisms and the nature of urban theory. *International Journal of Urban and Regional Research*. **40**.

Roy, A. (2009). The 21st-century metropolis: New geographies of theory. *Regional Studies*. **43**(6): 819–830.

Sangeet Natak Akademi (2019). *Durga Puja of Kolkata, Unesco*. Retrieved June 6, 2023, from http://sangeetnatak.gov.in/uploads/ICH-Inventory/029-Durga Puja.pdf.

Sarma, J. (1969). Puja associations in West Bengal. *The Journal of Asian Studies*. **28**: 579–594.

Schatzki, T. R. (1996). *Social practices: A Wittgensteinian approach to human activity and the social*. Cambridge University Press.

Sen, M. (2022). Between religion and politics: The political deification of Mahishasur. *Religion*. **52**(4): 616–636.

Sen, M. (2022). From demon to deity: Forging a new iconography for Mahishasur. *Journal of Material Culture*: 13591835221116708.

Shove, E. (2016). Matters of practice. *The Nexus of Practices: Connections, Constellations, Practitioners*. London and New York, Routledge: 155–168.

Shove, E. (2017). *Practice Theory Methodologies Do Not Exist—Practice Theory Methodologies*. Retrieved June 6, 2023, from https://practicetheorymethodologies.wordpress.com/2017/02/15/elizabeth-shovepractice-theory-methodologies-do-not-exist/.

Shove, E., M. Pantzar and M. Watson (2012). The dynamics of social practice. *Everyday Life and How It Changes*. London, SAGE Publications: 1–19.

Simmons, C., M. Sen and H. Rodrigues (2018). *Nine Nights of the Goddess: The Navaratri Festival in South Asia*. New York, SUNY Press.

Simone, A. (2019). Contests over value: From the informal to the popular. *Urban Studies*, SAGE. **56**.

Singh, T. (2018). Durga Puja pandals of Kolkata 2016: The Heritage and the design. *International Journal of Social Science and Humanity*. **8**(6): 155–159.

Sutton-Brown, C.A. (2014). Photovoice: A methodological guide. *Photography and Culture*. **7**(2): 169–185.

Truelove, Y. and N. Cornea (2020). Rethinking urban environmental and infrastructural governance in the everyday: Perspectives from and of the global South. *Environment and Planning C: Politics and Space*. **39**(2): 231–246.

UNESCO (2019). *Files 2021 under Process—Intangible Heritage—Culture Sector*. Retrieved June 6, 2023, from https://ich.unesco.org/en/files-2021-under-process-01119.

Wang, C. and M.A. Burris (1997). Photovoice: Concept, methodology, and use for participatory needs assessment. *Health Education & Behavior*. **24**(3): 369–387.

Warren, S. and P. Jones (2016). *Creative Economies, Creative Communities: Rethinking Place, Policy and Practice*. London and New York, Routledge.

Watson, V. (2009). Seeing from the South: Refocusing urban planning on the globe's central urban issues. *Urban Studies*. **46**(11): 2259–2275.

2 Crafts and practitioners

Idol-crafting practice and sustainability

Idol-making practices have evolved over time to fit contemporary festivities. However, the timing of completing and delivering is crucial and strictly governed by the Bengali calendar. In relation to this, an interviewee commented,

> Our profession is time-dependent; you have to deliver the idol to the customer before or on the day of the festival. . . . An idol becomes useless (and immersed in water) after the day of the festival. So, you see, timing is the essence.
>
> (Interview, Kumartuli 2018)

Here, I first disentangle the meanings of idol-crafting and idol-making practice. The former involves the process of sculpting and decorating the idol (presented in Figure 2.1), while idol-making practices encompass all the processes related to idol-crafting, from the commissioning through the procurement of materials to the disbursal of the finished idols from crafting workshops in Kumartuli and elsewhere in Kolkata. Various anthropological accounts have discussed crafting or craftmaking related to the clay idol sculpting and imagery in Bengal and the implications of this craft as a traditional artefact (Chakravarti 1985; Sen 2015; Agnihotri 2017). Idol-making practices in Kumartuli involve a multi-layered network of actors and stakeholders. I also argue here that the autumnal festivals of Durga Puja, followed by *Lakshmi* and *Kali* Puja, primarily constitute a majority of idol-making practices in Kumartuli. Other seasonal rituals related to idol-crafting have been historically less substantial. Table 1.1 outlines the different festivals celebrated in Bengal and eastern India, which necessitate crafting idols for worship. The approximate preparation time shows the engagement and employment of the *kumars* with time overlaps. Seasonal migration of labour from the hinterland is a prerequisite for the more prominent festivals in autumn. '*Rathayatra*', celebrated in the early monsoon, conventionally marks the beginning of the idol-making process.

Due to the changing traditional seasonality and a growing international clientele, most idol-crafting activities start earlier than that yesteryears, which disrupts the overall seasonal pattern and requires adjustment in the practices. The sales of

DOI: 10.4324/9781003341222-2

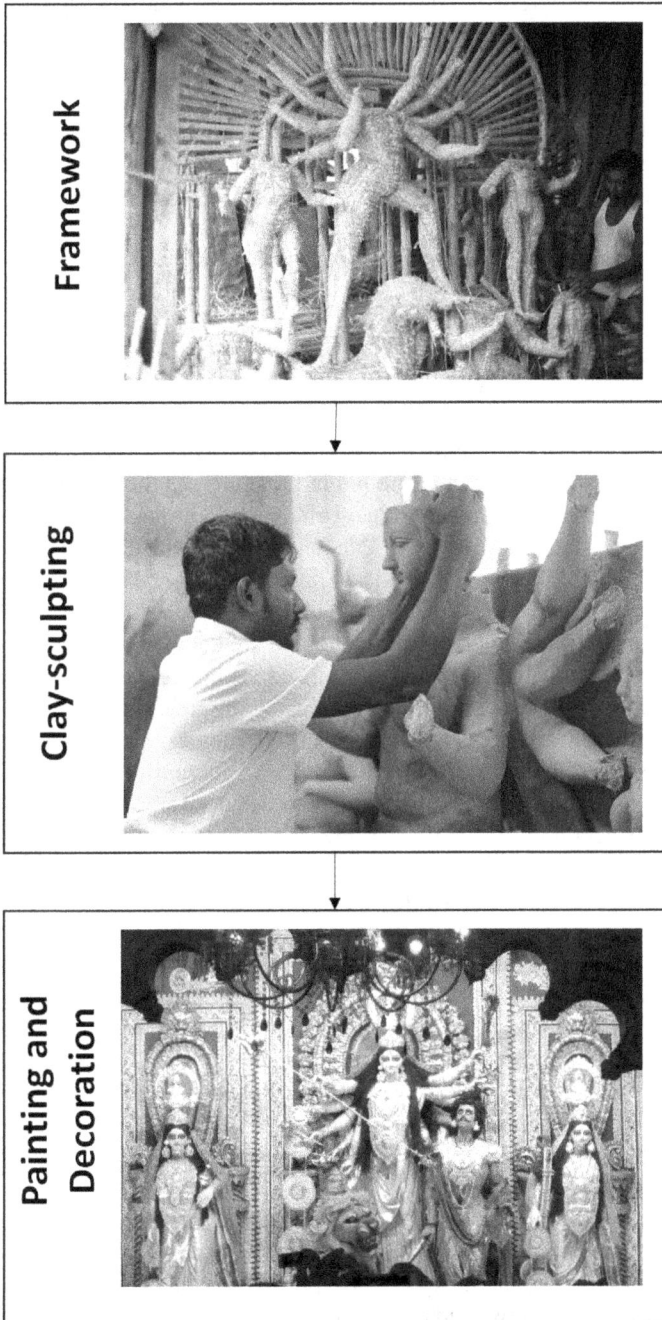

Figure 2.1 The process of idol-crafting: the three primary stages, including framework construction, clay sculpting, and painting and decoration

idols from the other festivals in winter and spring provide the capital for sourcing raw materials and labour payments for the Durga Puja production phase. This entire annual process involving sourcing of material and labour to spatial rearrangements to fit idol-crafting and delivery of the idols to customers on time is part of the idol-making practices based in Kumartuli. Also, this includes the immersion of the idols into the river and subsequent picking up of the residual bamboo or wooden framework for the following year's idols. Moreover, idol-making practices are closely related to the interconnected practices of idol worship and festivities.

I have structured this section using supporting materials from different literature as well as the lived experiences of the *kumars*. Through their photographs, the research participants[1] have illustrated materials and their spaces of practice. In addition to the secondary data and the interviews, reflections from the photo-study participants' perspectives were essential to construct the elements of materials and technologies of the practice. This section narrates the evolution of idol-making practices, and how specific know-how of idol-crafting has adapted over time. Here, I present both traditional practices found in literature and the contemporary styles of idol-making with changing consumer behaviour. Idol-making used to be a more household-based business. However, due to the rising demand for idols beyond the festivities of Durga Puja and other local Hindu festivals, the *kumars* hire seasonal workers to carry out the extensive workload.

The initial requirements for crafting the idol are bamboo slats, wood, and nails for the joinery. The river Hooghly has played an important part in transporting piles of bamboo downstream to Kolkata. Stacks of green bamboo are floated in the river from the rural areas towards the north. They are received at the Kumartuli *ghat* (waterfront embankment) by the local *mahajans* (merchant or supplier) (Figure 2.2). Bamboo naturally floats on the water. Transporting bamboo this way, through the river downstream, reduces the cost of porting and adds longevity to it. The water cures the green bamboo while increasing its strength and changing its colour to brown (interview: *kumars* and a supplier, Kumartuli 2017). The bamboo must be picked and carried to the embankments manually.

For the structural stability of clay idols, a bamboo and wood framework is constructed, which is called *kathamo* (frame). Idol-makers ritually worship the *kathamo* before the rest of the crafting process begins. As *kathamo* underpins the structure of the idol, this ritualistic worship is an essential part of the practice that *kumars* have traditionally followed through generations. Customarily, the *kathamo puja* marked the beginning of the idol-making process. *Kumars*, on the day of the kathamo puja, receive a symbolic advance payment towards the ordered idols of the client.

Kumars then start sculpting the form of the idol with straw. The straw is bound tightly with jute strings. After the essential structure of the idol is prepared using the straw, the first coat of sticky clay is applied over the straw-core (Figures 2.2 and 2.3). Earlier, the straw used to come through the river in boats from the adjoining rural agricultural fields. Due to urbanisation and decreased farming around the Kolkata Metropolitan Area, the straw now comes from rural areas in Bardhaman and Medinipur districts, a few hundred kilometres away. Straw bales are loaded

Figure 2.2 Top left: stacks of bamboo and clay-carrying boat ported from the riverfront *ghats*; top right: clay unloaded on the *ghat* (*source*: author, 2018). Bottom left: a resident *kumar*, sculpting straw framework sitting at the threshold of a workshop in Kumartuli; bottom right: a *kumar* sculpting a clay idol on the street (*source*: R7)

on lorries and arrive at the nearby Bagbazar area very early in the morning. '*At 4.30 AM in the morning, there are lesser traffic congestion and restrictions on Kolkata roads*' (interview: Kumartuli 2017). Once the straw is unloaded, they have to be bought by the *kumars* from the respective *mahajans*. During the interviews, *kumars* complained about the rising prices of straw and the lack of good-quality straw in monsoon because of rotting issues, reiterated by *kumars* several times during interviews. Some complain about the 'monopoly' business in Bagbazar by a particular *mahajan* and the lack of competitive prices. As a result, some *kumars* buy from elsewhere, farther away, adding to the cost of transportation (interview: Kumartuli 2018). The lack of dry storage space within the workshops of *kumars* adds to the problem, they said. A five-foot-high stack of straw in a workshop, costing about 600 rupees (about six GBP), only lasts a few days in the early months of preparation, mainly during the monsoon (interview: Kumartuli 2018). There is neither space nor favourable conditions to store the straw for longer than a week, so they have to purchase the straw on a regular basis.

The bamboo and the clay-loaded boats are an '*everyday sight*' of the Kumartuli *ghat*. These have been reflected in the photographs (Figures 2.2 and 2.3) of residents and an 'image' highlighted in essays on Kumartuli (Agnihotri 2017). The clay trade employs a number of merchants for the supply of raw materials to the idol-makers. During the preparation season of the idol, between the months of June and September, boats loaded with clay reach the riverfront *ghats* of Kumartuli. The *matir nouka* (clay-loaded boats) come in during high tides and are anchored by the embankment. The clay is unloaded and piled on the riverfront when the water recedes during the next low tide. Alternatively, due to the erosion of the riversides and restrictions of digging in populated areas, sometimes the clay comes from agricultural fields on lorries. *Kumars* buy clay directly from the merchant and carry it through the inner alleyways to be stored in their workshop or on the streets. The seasonal migrant workers employed by the *kumars* usually carry clay and straw from the *ghats* to the workshop.

Two types of clay are used in the idol-crafting process; a stickier variety, *entel-mati*, and a darker variety, *kala-mati* (Bean 2011). The river has generally been the source of procurement of clay. Originally, clay was excavated from the river-bed during low tides. A different variety of clay, much sandier in composition and locally called *bele-mati*, is used to make the finer details of the idol's face, fingers, and toes. Clay bought by *kumars* must be carefully sorted and prepared for application by checking for pieces of stones, brick, or other larger particles. Seasonal workers use their hands, feet, and spades for this purpose (Figure 2.3). Rice glue and water is mixed proportionately to make a firmer mixture with the clay to make it last longer.

From an ecological sustainability perspective, all the materials used in idol-making have always been in 'abundant supplies' in rural Bengal because paddy is the most cultivated crop in Bengal (interview: Kumartuli 2017, 2018). The straw and rice from paddy are directly used in idol-crafting. Clay, rice, and straw, the central part of an idol, disintegrate into the river water with time and is connected to '*vegetation, cultivation . . . with a capacity to sustain life*' (Bean 2011, pp. 608). An unfired clay idol for worship is, therefore, according to the Hindu tradition, meant to be immersed in water after the ritual; fired clay (*terracotta*) or a more lasting material is not a sustainable worship alternative. However, traditional clay idols are continually being replaced by contemporary artistic renditions in the current competitive practice of idol-making based on the clients' needs and demands. The *kumars* believe that having a clay idol for worship is essential. Even if a client commissions an idol for exhibitions and competitions, for worship and performing rituals, a mini clay idol will still be prepared in addition to the showcasing version (interview: Kumartuli 2017).

Fine white pieces of cotton cloth, woven by the weavers (*tanti*) in the traditional looms of Bengal, are firmly placed in pieces over the idol to add definition. A layer of white clay (*khari-mati*), found in the western Birbhum district of Bengal, is applied as a basecoat on top of the cloth-layer before colouring the idol. The pigments traditionally used for colouring the idols were naturally sourced and water-based. Previously, all the items for paintings were prepared from natural sources by the *kumar* family members at home; they usually buy pigments now (interview: Kumartuli 2017). These pigments are chemical compounds and look much brighter when applied compared to natural dyes. *Kumars* now opt for lead-free alternatives as much as possible, but these are quite expensive, and clients are hesitant to pay

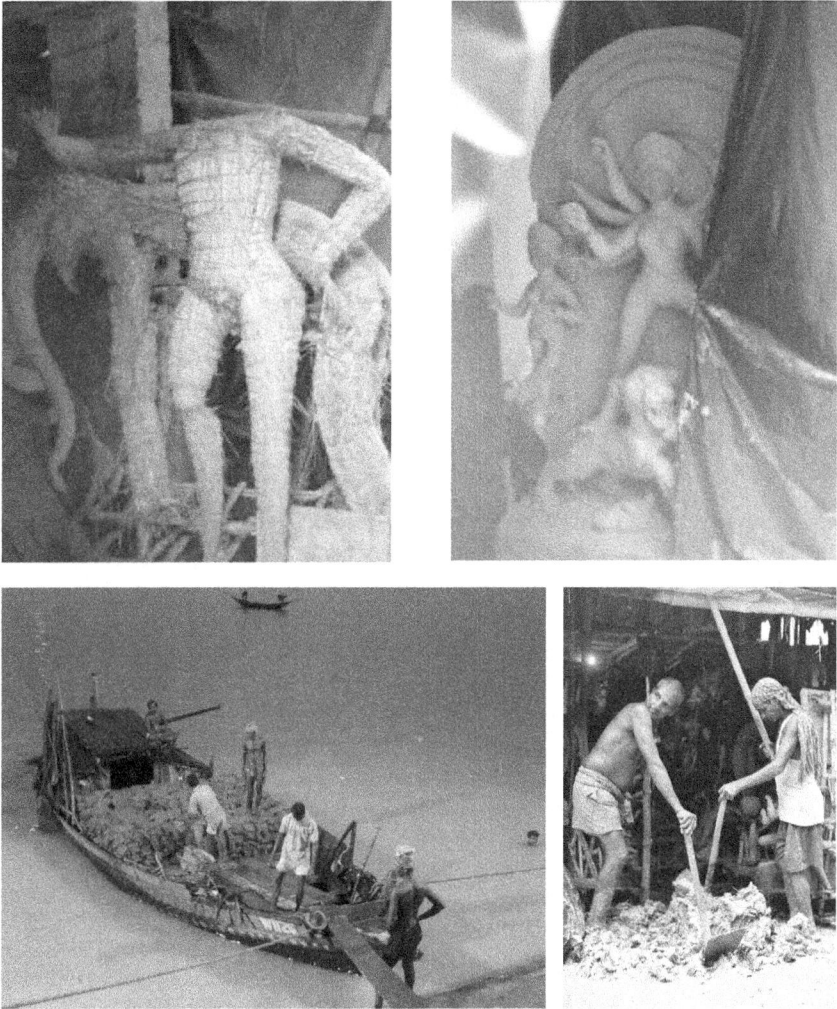

Figure 2.3 Top left: the initial straw framework of idols; top right: frame after clay sculpting, covered partly by plastic sheet from the weather. Bottom left: seasonal migrant workers unloading clay from a boat; bottom right: seasonal migrant workers preparing clay on the street

the difference. Hence, most idols are painted with chemical-heavy dyes, marking a massive shift from the traditional sustainable materiality.

The final ritual of idol-crafting is the painting of the eyes which requires maximum concentration by the *kumars* and is completed in privacy. The most experienced *kumar* in the household is engaged in painting the pupils of the eyes of the female deity. Historically, the most experienced *kumar* earned acclaim among his customers for the finesse of the eyes, locally believed to be the idol's most striking feature. *Kumars* started to become more individualistic and began developing their styles of painting the eyes, slightly different from their peers. Styles evolved in terms of the shape of the eye, the position of the pupils, the colour of the face of the idols, and the style of the ornaments used, making each style an individualised signature art form created by *kumars*. These acclaimed *kumars* of Kumartuli, along with their family, employees, and trainee *kumars* and ancillary networks, formed their brands among the clientele. Kumartuli earned international acclaim during the later stage of the nineteenth century. During the Great Exhibitions of London in 1851, some installations of the artisans from Kumartuli and Krishnanagar were exhibited (Bean 2011). Some artists were commissioned to prepare clay portraits of British dignitaries in Kolkata. Surprisingly, despite all the acclaim, most idol-makers continued to dwell and work in the marginal settlement of Kumartuli *basti* in the northern fringe of Kolkata.

Kumbhakar caste relates to pottery

In this somewhat simplistic narrative of the caste system in Bengal, I present the caste-based relations of the *Kumar* profession, and how some Bengal potters became idol-crafting practitioners to unpack the history of the settlement in Kumartuli. *Jati* (loosely translating to caste) within the Hindu religious and societal practices is conventionally based on the occupation of a person and inherited through generations (generally a man due to patrilineal relations). Anthropological studies reflect the complexity and prejudices associated with India's caste system, particularly on the caste hierarchies in Bengal (Sen 2015; Heierstad 2017). The hierarchical system that existed in Indian society was designated as the *Caste System* during the European Colonial period in order to rationalise the rigid two-tier social and occupational divisions. Within the two-tier caste system are hierarchies of the *Varna* (mostly social) and the *Jati* (occupational). The social class or the *Varna* governed the social status of a person, the most superior being the *Brahmin* (priest) and *Kshatriya* or *Kayastha* (warrior) castes, followed by the *Vaishya* or *Baniya* (merchants) and the *Sudra* (working, even menial) castes. The traditional caste division is usually reflected in a person's family name, and one was obliged to take up his (mainly men, as it is primarily a patrilineal society) forefathers' profession. Like many other regional variations and heterogeneity within India, among the Hindus in Bengal, the castes were also divided into sub-castes.

In Bengal, caste distribution is less emphasised in the *Varna* system's social hierarchies, comprising only *Brahmin* and *Sudra Varnas*. The *Jati* system has a greater sub-divisional hierarchy. The *Sudras* are subdivided into at least 14 such *Jatis*, including *Malakar* or *Mali* (garland maker and florist caste), *Karmakar* or *Kamar* (metal worker caste), *Tontubay* or *Tanti* (weaver caste), *Modak* or *Moyra* (confectioner caste), *Kumbhakar* or *Kumar* (potter caste) among others (Bose 1958; Heierstad 2017). The *Kumbhakar* or the potters who handled clay, adapted their practice to craft clay dolls and clay idols (Sen 2015). Here, I mention only the *Jatis* traditionally involved in the production of religious items or services. Interestingly, as mentioned earlier, all the labour-intensive support required to perform a religious ritual is usually received from the castes belonging to the significantly lower strata of the Bengali Hindu society. Except for performing the ritual itself, which is the sole duty of the *Brahmin* caste (priest), every other duty, including idol-making to preparing confectionery and flower garlands as an offering to the idol, to weaving and metal foundry, is a profession typically identifying with the socially and economically lower and marginalised castes. The contributions of these lower castes in the traditional labour market are indubitable. Yet, they remained in substantially dominant social and economic relations due to the imposing hierarchical nature of the caste system. Two other *Jatis* integral to Bengali Hindu social system, *Baidyas* and *Kayasthas*, historically belong to the non-*Brahmin* caste. Still, they are considered much higher in social position than the *Sudra Jatis*. They are mainly the landholding and professional castes such as doctors and teachers. The *Brahmin, Baidya*, and *Kayastha* were socially and economically more privileged 'upper' castes than the historically oppressed *Sudra* castes. This hierarchical system continues to exist and dominate the unequal social relations in postcolonial India.

From here on, I have used *kumars* and artists interchangeably because in Bengali, idol-makers are called *mritshilpi*, meaning clay artists (Sen 2015). To a modern artist, caste is a brand they carry and comfortably use as part of their heritage. In fact, *Kumars* prefer their caste-based professional reference and proudly identify themselves with their caste-based titles (surname) as Pal, alternatively, Paul (interview: Kumartuli 2017).

The caste-based potters' *para*

In the northern fringes of the then Kolkata (originally the village of Sutanuti),[2] the settlement of a community of potters can be traced since the late seventeenth century. This settlement might have been because Sutanuti was developing as an influential market area close to the river trading hub and slowly becoming the epicentre of the native settlement at the beginning of the European trade scene. The people belonging to the *Kumbhakar* (*Kumar* caste) perhaps settled in Kumartuli due to the traditional habitation concept of *para*[3] prevalent in the villages of Bengal. In rural Bengal, *para* is a neighbourhood often restricted to a single social and occupational caste or *Jati*. Several such *paras*, constituting families of kin and each with a distinctive boundary, collectively formed a village. Similarly, in early Kolkata, formed out of the amalgamation of the three villages, *paras* were localities of families with a similar occupation or mainly migrants from the same village or area circle (Sinha and Bhattacharya 1969; Roy 2012).

Kumartuli literally translates to 'neighbourhood of potters' in Bengali and was used as an administrative area to designate the *para* of potters. In Bengali, *Tola* is an area of a homogeneous population group, usually smaller than a *para*. *Tuli* is a smaller *Tola* (Archer 2000; Heierstad 2017). In the earliest years of Kolkata, the names of areas continued to be taken up from the already existing village locations of markets (*bazaars*), ponds (*pukur*), *tolas*, and *tulis*. Hence, the area's name, Kumartuli, suggests the existence of a small neighbourhood of potters in Sutanuti village before the British arrived. Most large Bengali villages have a potters' neighbourhood, and records show the existence of a potters' neighbourhood in Gobindapur (another one of the constituent villages of present Kolkata). However, only Kumartuli in Sutanuti flourished as the idol-makers' cluster, whereas the 'Kumartuli' in Gobindapur remained a much smaller area and was lost to the growth and urbanisation pressures of colonial Kolkata.

In Kolkata, the earliest administrative divisions made during the colonial period were around a *thana* (police station) in the centre. *Thanas* were the smallest units of the division, the areas around it being the administrative boundary. By the early 1700s, Kumartuli was part of a popular *thana* area for the Bengali *babu* class (aristocrats). Street names in the Kumartuli area are reminiscent of the various *babu*s residing in the locality in the early to mid-eighteenth century, the most notable being Banamali Sarkar and Nandaram Sen (Figure 2.4). The surrounding areas of Kumartuli in northern Kolkata were the constituents of the native or 'black' town during the colonial period, dotted with the *bonedi bari*[4] and other palatial mansions of the *babu*s and the modest houses of Bengali *madhyabitta* (middle class).

Although Kumartuli's history has not been widely documented, it can be assumed that potters migrated from Krishnanagar (a small town in the administrative district of Nadia, a few hundred kilometres north of Kolkata, shown in Figure 2.5) and adjoining areas to the already thriving areas of the then Kolkata in search of better livelihoods (Agnihotri 2017; Heierstad 2017). Additionally, after the beginning of the popular culture of celebrating the autumnal Durga Puja, potters from Krishnanagar were brought to be settled in Kumartuli near the Shobhabazar *Rajbari* (palace or mansion) and other such *bonedi baris*. Bean notes that the nearby presence of the elites, who commissioned the idols, nurtured this artistic community to grow in number. In order to respond to the rising demand for

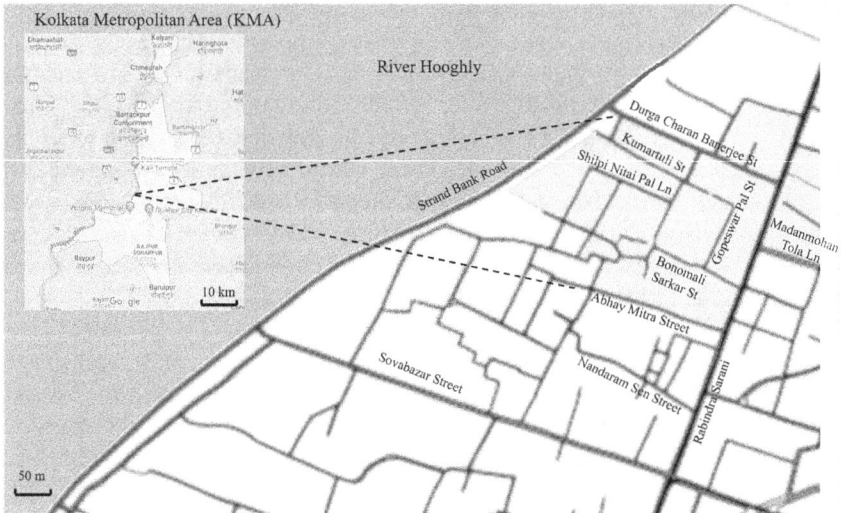

Figure 2.4 Map delineating the extent of the Kumartuli potters' neighbourhood drawn with respect to the location of Kumartuli in the Kolkata Metropolitan Area (KMA)

KRISHNANAGAR
(Clay-idol making)

Seasonal migration of labour to other parts of India

From BANGLADESH

GHURNI
(Shola carving)

Shola carving

Migration of potters

SHANTIPUR

Continuous exchange of labour and raw materials

Migration of potters

Clay-idols

Small clay idols

RURAL BARDHAMAN

RANAGHAT

idol Making Hub

RURAL NADIA

Supply of Shola

CHINSURA

(Terracotta products, lighting industry)
CHANDANNAGAR

Seasonal porters

Seasonal idols

Tourist

Shola Decoration

Supply of bamboo (through river)

Seasonal migration of potters

HOWRAH   KUMARTULI
Kolkata

RURAL HOWRAH
(Small cluster of allied industries)

North KOLKATA Trading Hub

Migration of potters from BANGLADESH (1940-1960)

Supply of clay

Wholesale of allied products to other parts of India

Pandal decorations

ULUBERIA

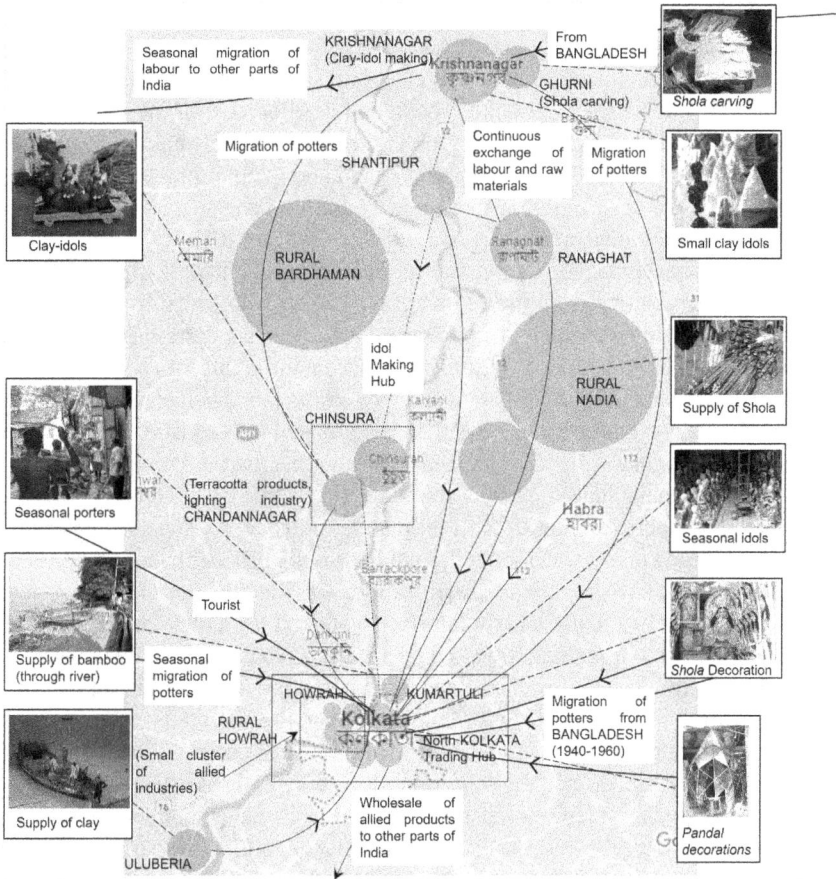

Figure 2.5 Historical and contemporary supply chain map

Source: author

these idols, the potters perhaps brought in extended families for help from the rural areas (Bean 2011).

Durga Puja was first recorded formally to have been ritually worshipped by Maharaja Krishnachandra Roy's ancestor in the capital Krishnanagar in Nadia in the early seventeenth century (possibly c. 1606) (Chaliha and Gupta 1990; Bhattacharya 2007). The festival was further popularised by Maharaja Krishnachandra Roy when he started the worship of '*Shakti*: the cosmic force manifested in the forms of a goddess' (Bean 2011) on unfired clay idols to be immersed at the end of the religious worship. The Sabarna Roy Choudhuri family of Barisha, in the southern fringe of present Kolkata, were the *jamidar* (landlord or landholding families, also spelt as *Zamindar*) who owned the villages of Gobindapur and Kalikata (two of the three constituent villages of colonial Calcutta), before Job Charnock, an administrator at the English East India Company, bought them for business in c.

1690. This *jamidar* household records their Durga Puja celebration since c. 1610, perhaps the first in the Kolkata area (Chaliha and Gupta 1990).

In other words, the elites required to celebrate Durga Puja to assert their position of power in the society, and the growth and development of the idol-makers' cluster were a direct outcome of this growing demand for idol-making in Kolkata (Bhattacharya 2007). The complexity of the land and the building ownership was the direct outcome of bringing in more potters from Krishnanagar and continued patronage to settle down in the *Zamindar*-owned land in Kolkata. The complex tenure agreements existing in the Kumartuli *basti* were an outcome of this colonial tenancy system resulting from the demand for ritualistic idol worship.

Therefore, the initial acclaim of Kumartuli emerged as a caste-based brand by the *Kumars* who migrated from Krishnanagar during the eighteenth and nineteenth centuries. Krishnanagar-based pottery and clay crafts are considered the epitome of crafting in Bengal, and the place is the brand related to practising craftspeople. The potters, particularly in Krishnanagar, due to the growing demand for idols, both large and small, started carving the idols out of soft clay (Ghosh 2012). To trace the origin and evolution of the idol-making practice, I visited Krishnanagar to understand the current situation. Clay idols or dolls of Krishnanagar are much smaller in size, unlike the human-scale or more massive idols of Gods and Goddesses of Kumartuli, and are mostly figurines made of clay and painted. These are sold for decorative purposes to a wide range of users. This small-scale industry of Krishnanagar remains an important household-based industry and attracts tourists to Krishnanagar and the adjoining Ghurni. Similar to the potters of Kumartuli, Krishnanagar potters are also sculptors of the same *kumar* caste and are employed in this trade on a generational basis (interview: *Kumar* in Krishnanagar 2017). Based on multiple interviews in Krishnanagar and Kolkata, the potters' community of Krishnanagar were the best clay sculptors of Bengal during the early eighteenth and nineteenth centuries. They crafted clay dolls of much smaller sizes that were not meant for religious purposes.

While Kumartuli evolved as an idol-making community based on local demand in Kolkata, the practices in Krishnanagar evolved to become one of India's famous clay-doll-producing centres. Both communities of practice originally sculpted unfired clay idols following similar elements, such as materials and technical know-how related to the practice, but both diverged and evolved on the basis of their relational geographies and production demands. Both centres emerged differentially on their sociocultural and political relations and political pressures over time. It should also be noted that other potters or idol-crafting clusters in Kolkata and adjoining areas have developed as well, such as in Kalighat, Jadavpur in the southern part, and Baranagar in the northern part, and more recently along the Circular and Keshtopur Canals in the northern fringes of the city. These clusters are smaller in size, and their histories cannot be traced back as early as Kumartuli. Hence, Kumartuli remains the oldest, largest, and, due to the recent redevelopment initiative, the most prominent of all idol-crafting clusters in Bengal.

Extensive migration from Krishnanagar to other parts of Bengal, like Kolkata and Chandannagar, took place in the late nineteenth and early twentieth centuries when idol-making became a vital part of the religious and cultural ingredients of city life. Drawing from interviews, most Kumartuli potters link their roots to Krishnanagar; however, my interviewees were unsure as to when exactly their families migrated or what caused them to migrate. The initial migration of potters from Krishnanagar was not enough to suffice the needs of idol-making in Kolkata and the adjoining areas. Potters migrated to Kumartuli over a long period until the 1950s and 1960s. Also, every year a large number of seasonal workers migrate to Kolkata for employment within the idol-making industry over a few months and return home. Seasonal migrants include members of extended families of the Kumartuli potters, as well as agricultural workers from the rural areas opening up the wider geographies of the idol-making practice. Through interviews and ethnographic mapping, I have traced this information on a map to present the flows of goods, people, and capital in the idol-crafting industry in Kolkata in Figure 2.5.

After the partition of Bengal and India's independence, a large number of potters from Bangladesh also migrated to Kumartuli. Families of potters escaped communal riots in Bangladesh to move to Kolkata, like many other communities. They settled in Kumartuli because they had earlier acquaintances here that belonged to the same caste, and finding a job as a potter or idol-maker seemed easier in Kumartuli (interview, Kumartuli 2017). This community of additional migrant potters, while living in Bangladesh, were occupied with both pottery and idol-making, idol-making being a seasonal business, whereas pottery was year-round (Agnihotri 2017). Settling in Kumartuli meant they could be employed in idol-making practices while also producing clay utensils. Figure 2.5 traces potters' historical geographies and seasonal employee relations that shaped the present Kumartuli.

Some *kumars* of Kumartuli believe that it was due to the decrease in demand for clay utensils that their ancestors engaged fully in idol-making, while others believe that their ancestors migrated to Kumartuli to fulfil the demand for idols for rituals of kings and *babu*s (interview: Kumartuli 2017). The potters also used their wheels to make shells for clay (usually *terracotta*) fireworks like *tubri*[5] and *uron-tubri*,[6] both displayed during festivals. This tradition was quite common among young people until the government banned the use of flying fireworks displays due to the safety of the Durga Puja *pandals* in the late 1990s (interview: Kumartuli 2017). In order to maintain their livelihoods, the *kumars* in Kumartuli have altered their commodities to serve the people of Kolkata with essential items for worship over the years, including items made on the potters' wheel, like painted clay utensils, pots, *diyas*, to smaller clay figurines of Hindu deities (interview: Kumartuli 2017). A few families still prepare these items in small numbers, mainly the women and very old *kumars* of the household. Therefore, pottery, small clay dolls, and terracotta crafting, which were the original occupation of the *kumars* migrating from Krishnanagar and Bangladesh, are now subsidiary businesses of an idol-making household in Kumartuli (interview: Kumartuli 2017). The relation that connects Krishnanagar, some 120 kilometres north of Kolkata, is the practice of idol-making. The practice that once emerged in Krishnanagar or the Bengal district of Nadia has evolved and hugely altered, still carried on by the *kumars* of Kumartuli primarily supported by the geographies of power through the socio-political relations of the emerging *babu* class that the then colonial Calcutta was evolving to become. The timeline (detailed in Table 2.1) briefly tabulates the history of the Kumartuli neighbourhood.

Layered on a map of the Hooghly basin, Figure 2.5 shows the historical and contemporary supply chain map developed through fieldnotes and interviews in

Table 2.1 Timeline of events relating to the idol-making community of Kumartuli

Period	Date/Period of the Event	Event Description	Implications on the Kumartuli Area
Up to 1757 (before the British Imperial era)	Before 1690/1698	The existence of Kumartuli as a small potter's neighbourhood in the then village of Sutanuti	Resulted in the designated name of the Kumartuli area in Kolkata
	1717	Establishment of Kumartuli Thana (administrative division) and a few *babu*-families settled	A rise in the number of migrant potters from Krishnanagar
	1757	Change of governance of Bengal	First population data recorded shows a few potter families
1757–1947 (colonial governance India)	1761–1790	First *barowari* puja	The rise in demand for idols
	1851–1854	Great Exhibitions in London	The popularity of certain sculptors among the Europeans resulting in rising business
	1913	First *Sarbojonin* Durga Puja, supported by the Indian National Congress	A rise in demand for *swadesi* items resulting in a resurgence of locally sourced ornaments and dresses for the idols
	1931	Kumartuli *Sarbojonin* Durga Puja by the then Congress President Subhas Chandra Bose	First *Sarbojonin* puja in the potters' area, which continues till date
1947–1972 (post-independence India and Bangladesh Liberation War)	1947	Partition of Bengal and India	Migration of potters from Bangladesh to Kumartuli, resulting in a huge increase in the population of Kumartuli
	1948	Thika Tenancy Act	Tenants' rights reserved for the first time

Period	Date/Period of the Event	Event Description	Implications on the Kumartuli Area
1977–2011 (Left Front Government in West Bengal)	1977	Change in local governance	Kumartuli workers' cooperative setup
	1981	Slum Improvement (Clearance and Eviction) Act	Improvement of basic services like electricity and water supply in Kumartuli
	1980s	Local government-sponsored formal competitions with associated prizes among *Sarbojonin* Puja committees	A gradual change like competition among the potters; few potters gained more popularity than others
	2007–2011	Basic Services to the Urban Poor (BSUP) Jawaharlal Nehru National Urban Renewal Mission (JNNURM)	Improvement on roads, lanes, and sewerage
	2008–2011	Redevelopment proposal by KMDA and survey of demographic and housing conditions in the neighbourhood	Collective resistance backed by the political opposition party
2011 to present (Trinamool Congress Government)	2011	The shift in governance policies	The redevelopment project was turned entirely down
	2013- present	State-promoted tourism and cultural industries' policy intervention	Annual immersion rally and increased tourist footfall in Kumartuli
2019	2019–2020	Nomination for listing Durga Puja in UNESCO's Intangible Cultural Heritage list	Yet to be understood fully; however, getting international attention is inevitable

November 2017. It shows the migration of people and the flow of materials through the river influencing the development of the idol-making cluster in Kolkata (c. 1800–1950s). Kumartuli is at north of Kolkata, and Krishnanagar is about 120 km north along the river. The layers and lines on this map are diagrammatic representations; grey circles are places directly associated with the idol-making industry and the allied crafts. The circles are not to scale; the lines show linkages. The river Hooghly, being an integral part of the everyday life of the residents of Kumartuli, is not only the source of transport of bamboo and clay but historically was the means of distribution of the idols as well. Also, the supply chain has slowly developed over centuries and continues to be mostly dependent on the idol-making cluster in Kumartuli and nearby Baranagar (just outside Kolkata's municipal periphery). This map signals the more comprehensive network along with relevant images located across multiple geographies that have shaped the practices in the Kumartuli and continue to facilitate the constant organisation of relations within a sizeable and thriving informal sector industry.

From the constructed supply chain map, I argue that most practitioners are located in and around Kumartuli mainly on the basis of the need of their practices. The map and the earlier materiality discussion suggest that there is a need to be based in Kumartuli to be at the centre of idol-production of Kolkata to procure raw materials and find a seasonal workforce and the need to be based in proximity of the socio-spatial fabric of an urban potters' community. Maybe this also was the reason for the Bangladeshi potters to come and settle down in Kumartuli. So other than kinship and caste-based preferences, there remained a place-based connection of certain spatial practices uniquely related to idol-making, and Kumartuli as a neighbourhood remained at the centre of all these interwoven practices.

From a practice theory perspective, the materiality of the practice is essential (Shove et al. 2012), as are the materials and the processes that lead to the production of the idols. Traced through mapping the local and global flows, the practice of idol-making may have been embedded in place due to the existence of the materials. Also, the sociocultural and political processes, such as the religious festivities of Durga Puja from the *babu*-family with the city's growth and other relations, have contributed vastly to the success of the idol-making practices. Additionally, the social interaction within a caste-homogeneous community may have contributed to constructing the place identity. The caste positionality of the *kumar* places him within the subaltern[7] groups that have been historically marginalised, lacking representation and voices of expression. Their only mode of expression, idol-crafting, was also dominated by the requirements and consumer demands of the landed gentry and the *babu* classes. However, the artists of Kumartuli are slowly beginning to receive the freedom of expression through their art installations in some Puja *pandals* and idols in Kolkata.

The participatory data suggests the idea that the sociocultural practices are shaped by the 'place' since the inception of the neighbourhood, meaning the residents are keen on holding on to their existing spaces. Participants have taken pictures of everyday places in their neighbourhood that they think are important and contribute to idol-making practices. These include pictures of the riverfront, temples, and street corners. Each picture has a background story with emotional meanings, sentiments, or memories, which the participant narrated during the follow-up interviews. Also, the photographs presented in this book with corresponding narratives are significant in constructing the meaning of the place with the practices that have evolved here.

From the supply-chain map and discussions with the local residents, the river Hooghly seems to be a crucial part of Kumartuli's historical growth and development narrative as Kolkata's idol-making hub. Participants have also pointed out how they feel that the river plays a role in everyday life of each resident in different ways, and how the existing multi-modal network supports the practices. However, the question remains whether the shifting practices relating to Durga Puja, both organically and triggered by the government's planning body, would successfully reshape the practices elsewhere, away from the existing resources available in Kumartuli. Also, the shift would affect the historical relations shaped by the place; would the practices realign with the new geographies?

Interwoven communities of practice

The demand for ancillary professions like pith carving and ornament-making started growing as the Durga Puja celebration gathered popularity outside the *babu* mansions. These supporting practices started from Kumartuli as a centre of production in Kolkata. Previously, these crafts were also popular in parts of Bengal, especially in the Nadia district. An abundant material growing in the water bodies in rural Bengal, pith from roots (*shola*) (shown in Figure 2.6), has been adorned as decorative paraphernalia in Bengali festivities (Agnihotri 2017). Ornaments, decorative features, and hats carved from *shola* were extensively used in Bengali Hindu rituals. The idols' ornaments (*sholar saj*) were prepared from carved *shola*. Alternatively, ornaments made from paper and metallic sheets (*daker saj*) adorn the idols. *Zari* is also a common alternative for the jewellery and embellishment of idols in recent years (interview: Kumartuli 2018). A tamarind-glue binder, again

Figure 2.6 Image of a shop exhibiting *shola* merchandise

a naturally sourced material, was extensively used for all adhesive purposes like fixing the idols' hair, jewellery, and dresses.

The *shola* crafts industry based in Kumartuli faces imminent threat due to the lack of availability of the material and decreasing number of artisans. On the contrary, an increase in the use of alternative synthetic, as well as *zari*, ornaments keep their businesses in the smaller stores running (interview: Kumartuli 2018). As Kumartuli grew, *malakars* (flower garland and *shola*-crafting caste) settled in the periphery of the idol-makers' cluster in the neighbourhood. Similarly, merchants selling other items like outfits set up shops in Kumartuli. A number of businesses in the nearby Barabazar, Kolkata's oldest trading hub, as well as a few shops in the Kumartuli area sell outfits and ornaments for the idol. These merchants are wholesale traders trading their items across eastern India and beyond. Every ancillary product required for finishing the idol for worship, like the ornaments of different types, the outfits, and even the *zari* used in the ornaments and garments, has a wide clientele outside Kumartuli and employs a sizeable number of workers. Hence, the practices allied to idol-making, although once interwoven and dependent on Kumartuli's businesses, have established economic relations beyond it and continue evolving and shifting simultaneously with idol-making practices.

During the peak autumnal festive season, when idols are traded in Kumartuli, the ancillary businesses contribute to supplying goods in smaller quantities for domestic rituals to local residents. These smaller shops sell *shola* items and decorative elements required for Hindu household festivities as well as *barowari* pujas (Figure 2.6). Other ancillary businesses in the Kumartuli area include the '*decorators*' industry: the bamboo merchants sell poles of bamboo, jute ropes, and sheets of plastic and canvas to erect *pandals* (marquee or makeshift pavilions) for housing the idols and pujas. The sales peak just before Durga Puja.

The neighbourhood's residents are also involved in an array of other allied practices unique to the neighbourhood. These smaller but important economic activities have also been reflected through the participatory photo-study. A resident (R7), a middle-aged lady, acknowledges the immense support the ancillary trades based in Kumartuli and adjoining areas provide to the idol-making practices. She said,

[B]ecause I have grown up here in Kumartuli, I understand that every other profession than idol-making forms an important part of the industry.

For example garment making, or the hosiery industry, is among the ancillary industries based in Kumartuli and employs a number of women, including women from *kumar* households. One of the seasonal supplies for idol-making, the hair made of jute fibres are ferried by the occasional seller who only comes to the Kumartuli neighbourhood and no other neighbourhoods of Kolkata (interview: Kumartuli 2018). According to the respondents, these two trades are unique to the Kumartuli neighbourhood and link directly to the idol-making industry. Supporting trades in Kumartuli reveal a unique multiplicity of practices that relate to the construction of the place while also shaping a larger socio-spatial landscape alongside dominant commercial practices of idol-making.

Social cohesion due to caste homogeneity over generations of residents brings about a sense of place that is singular only to the neighbourhood of Kumartuli. Also, a common and interlinked economic pursuit may have nurtured a further close-knit community. The interwoven practices demonstrate that idol-making and allied practices are more or less based in and around Kumartuli spatially. The most significant allied practices include *shola*-carving, dress and jewellery making for the idols, and *pandal* (marquee or exhibition pavilion) construction businesses. Besides the main festival of Durga Puja, idol-making practices in Kumartuli produce other seasonal commodities like smaller idols, statuettes, and sculptures around the year. The idol-crafting activities are interwoven with the everyday household practices of the resident potters of Kumartuli. All the smaller businesses, like the *shola*-carving and other ancillary practices linked to idol-making practices, are also produced in workshops attached to residences within the Kumartuli neighbourhood. Durga Puja, which was once the reason for this neighbourhood's development, is shifting and reshaping the practices based in Kumartuli.

Practices relating to Durga Puja that were once based in Kumartuli were *pandal*-making, rituals such as the first clay collection from the nearby red light area[8] (Chatterjee and Khullar 2004), storage[9] of idols during the preparation phase, immersions, and picking up of *kathamo* afterwards. The scale of all the operations has increased significantly and is slowly shifting away geographically from Kumartuli. The shifting of idol-making and related practices to Durga Puja, which was once crucial for the lives and livelihoods of the residents of Kumartuli, poses challenges to the idol-makers. Along with the related practitioners, *kumars* have to constantly reorganise their practices based in Kumartuli and elsewhere to accommodate the growing demands in the numbers and size of the idols. In doing so, some acclaimed *kumars* have permanently shifted their workshops outside the idol-making cluster but remained close to the resources available in Kumartuli, like the raw materials and access to transport infrastructure and amenities (interview: Kumartuli 2018). However, a large section of the *kumars* still retain their family workshops or a business premise at the centre of Kumartuli's idol-making cluster.

Emerging actors and shifts

With the increasing demand for the idol and the number of celebrations in Kolkata growing manifold, the potters in Kumartuli have become competitive and enterprising (Basu et al. 2013). In Kolkata and its neighbouring areas, the number of Durga Puja celebrations sum up to around 5,000, with an increasing number of licenses issued each year. The political interest and the nationalist movement manifested through the *Sarbojonin* Durga Puja during the independence movement in Kolkata have been shifted to local political leaders supporting their neighbourhood celebrations. *Sarbojonin* Durga Puja remains a democratic event, although politicians, party leaders, and ministers continue their patronage in the neighbourhood or constituency festivals to gain popularity, irrespective of their party affiliations among the general public and continuing their political campaigns (Roy 2003). Neighbourhoods commission highly priced idols from acclaimed idol-makers to participate in state-sponsored and industry-led competitions. Like *babu* families in the past, *barowari* Puja committees ritually perform the *kathamo puja* and mark the beginning of the year's Durga Puja celebration. Moreover, the Puja committee members say that they enjoy arranging the annual feast for the local community on that day to mark the beginning of the festivity; and they are determined to carry that tradition. However, on questioning ideas of traditions, Puja committees believe that the public now, with the rise in the number of *barowari* pujas, have many options to choose from and participate in the specific type of organisation they prefer. The Puja committees usually prefer to carry on with their own style to maximise participation and sponsorship.

Competition is not restricted only to the style and size of the idols. The more significant budget celebrations, mostly backed by political leaders, compete for the best pavilion (*pandal*) and the overall ambience of the locality during the festival. Judgements for these competitions are based on intricately and elaborately built *pandals* (Chattopadhyay 2019), lightings, the sound effects, though mainly the drummers (*dhakis*) and the idol being the centre of attraction (Figure 2.7). The bigger budget celebrations involve lakhs and crores of rupees (between thousand and hundred thousand pounds) (interview: Kolkata 2017).

In the current consumer-driven competitive market, idols are judged in terms of bigger sizes, more artistic adaptations, and meeting the relevant fashion trends.

Figure 2.7 Top—images of modern-style idols; bottom—images of different styles of *pandals*

For example some organisations compete for the best idol based on public choice. Posh neighbourhood Puja committees demand idol-makers to suit trending popular culture. Idol-makers sometimes have to alter the idol's faces or attire to resemble celebrities. The commodification of the idol has led fashion houses and high-street jewellers to sponsor and decorate some idols to advertise their range of products. Moreover, major corporate houses sometimes sponsor ambient concepts, themes, and even festive decor. The city is flooded with hoardings and flexes of advertisements leading up to the iconic Puja *pandals*.

One of the Durga Puja committees having the costliest budget in Kolkata, organised in the Sreebhumi Lake Town area in 2014, had all the five idols adorning jewellery made of precious gemstones like diamonds, emeralds, and rubies fitted in gold. The jewellery alone was valued at 10 crore[10] rupees (Indo-Asian News Service 2014). According to local newspapers, this showcase was a considerable step that significantly exceeded the gold jewellery embellishments from the idols of the previous year. An estimated 4 crore rupees (approximately 400,000 GBP) were spent on idols and embellishments in 2013 by the same Puja committee. However, this is not the only example of such extravagance in the festivities; gold and silver adornments of the idols are part of the exhibition that transcends ritualistic worship. With each passing year, the peaking popularity of this culturally vibrant festival and the organising committees being supported by politicians and corporate players, the celebration budget increases exponentially. Peer puja-organising committees increase their festival budgets through corporate sponsorships and invest in marketing and branding to promote their Pujas. The role of the *kumar* in this regard becomes insignificant, and the major highlight is the sponsoring jewellery chain or the corporate fashion house. The idol-makers in these festivities do not make it to the newspaper headlines, while the millions spent on the embellishments are in the spotlight. The spectators visiting the *pandal* on any evening *pandal*-hopping tour during the festival would perhaps not notice the name of the *kumar* printed at the base of the idol, the interviewee artists exclaim with a 'sigh' (interview: Kumartuli 2017).

I received mixed reactions when asked what share of the money spent on these celebrations goes to the idol-makers and the seasonal workers. With a sad undertone, some of the *kumars* mention that they get the minimal. While some wholly ignored the pricing of idols and wages; one of them, when questioned about it, said,

[Y]ou see, it is always dark just beneath the lamp, whereas it is supposed to light up its surroundings: meaning we craft idols to please the customers, but we cannot please ourselves and our families with the money we make.

(interview: Kumartuli 2018)

Every year, the price of raw materials increases. With even a tiny increase in all the items singularly, the cost of producing the finished idol increases massively. Nevertheless, customers always demand 'last year's price'. Many of the interviewees believe that if they deny negotiations with their existing customers, someone else in the neighbourhood would agree to a lower price, and, ultimately, he would lose important business (interview: Kumartuli 2018). Therefore, most idol-makers are bound to stick to competitive prices for the fear of losing customers. However, despite the lack of transparency in the fee structures, wages, and profit margins, it should be noted that artists continue to pursue their careers in idol-crafting, and each year more seasonal migrants join the workforce.

The idol-making industry has now transcended from being few-household idol preparations to a range of medium-sized to larger idol production and *pandal* decoration, including international commissions, presented in Figure 2.8. The approximate price of Durga idols in 2018 was about 15,000 to 30,000 INR (GBP 150–300) for small idols, about 50,000–100,000 INR (GBP 500–1,000) for medium idols, and anything above 100,000 INR to 1,000,000 INR (GBP 1,000–10,000) for larger idols. Additionally, the idols that are shipped abroad are much more expensive despite their smaller sizes. The primary idol-maker, or the *kumar*, has now modified his role as the principal advisor in the idol-making process, only doing the more specialised jobs such as face painting and mainly the eyes. They are the artists who conceive the idol's design and competitive style based on public demand. A Puja committee usually commissions the idol or the conceptual idea of the festival as early as January each year. Some higher-budget celebrations would commission

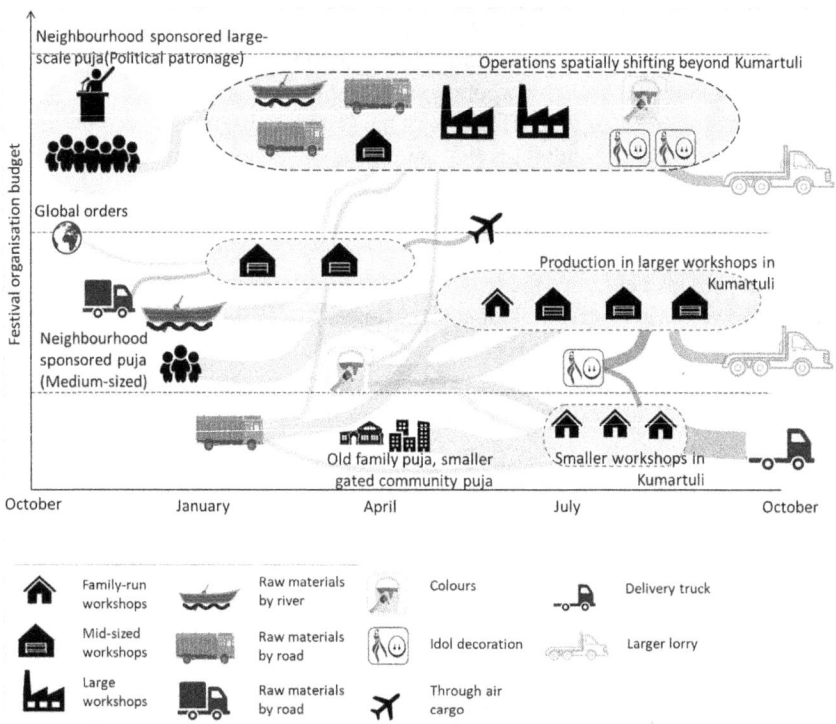

Figure 2.8 Idol commissioning and production diagram showing the flows and network of actors within the wider political economy of the idol-making industry that transcends beyond the Kumartuli neighbourhood

even earlier. The big Puja committees contract artists to develop the theme and conceptualise the spatial configurations involving the preparation of idols as well as the construction and management of the *pandals* (Guha-Thakurta 2015; Chatto-padhyay 2019). The entire process is 'time-consuming' and requires the knowledge of structural construction and interior design. A fine arts qualification is not essential but provides the knowledge and expertise to carry out these jobs (interview: Kumartuli 2018). These artists mostly have their own supply chain. However, some of them still rely on traditional suppliers, similar to the *kumars* of Kumartuli. The higher-budget festivals have sufficient scope for the advertisement of acknowledging the artists. They put up flexes and hoardings, promoting their Pujas and highlighting the artists' names. However, such emerging thematic presentations of the idol and the *pandal* are restricted to high-budget celebrations only and artists with a fine arts qualification. Artists are of differing opinions regarding this emerging trend. Some believe a diverse young group of artists with various caste backgrounds are 'definitely' required in the Durga Puja business. Others feel the need for skill training, awareness, and interest among the younger caste-based potters to continue their forefathers' profession; they think younger generations must learn idol-crafting first to move on to more creative renditions of the traditional craft form (interview: Kumartuli 2018).

Along with the local demand for Durga idols, Durga Puja has increased internationally among the Bengali-Hindu diaspora. Artists of Kumartuli have been shipping their 'light-weight' idols abroad. They market their commodities online through web pages and social media (Chakrabarti 2022). Tourism and revenue generation through the cultural industries and festivals, primarily through the Durga Puja of Kolkata, has been a priority of the local government, which assumed office in 2011 (Chattopadhyay 2012; Guha-Thakurta 2017; Chattopadhyay 2019). Recently, disciplines such as political science, history, and religious and political anthropology have explored intersections between religion and politics to conceptualise 'political deification' (Outlook Web Desk 2016; Sen and Nielsen 2022). In the contemporary political landscape of Bengal and India, the interconnectedness of politics and Durga Puja has become ever so prominent. The deity of Durga, in some cases, has emerged as a piece of political propaganda that suggests the Chief Minister Mamata Banerjee is the slayer of the demon '*Mahisasura*' represented by Modi and Amit Shah, Prime Minister and Home Minister of India (Sen and Nielsen 2022).

The local government organises an annual immersion rally for the celebrations receiving the highest accolades since 2016 (Das 2016). The Chief Minister of Bengal envisions this rally as a means to attract foreign tourists and facilitate a better tourism infrastructure for Kolkata. The idols that make their way into the rally are usually accompanied by the commissioning neighbourhood Puja committee and the most influential person (usually the political leader or a celebrity) associated with it. Artists complain that although their idols and concepts earn accolades, they are not fortunate enough to earn that fame or feature alongside their art (interview: Kumartuli 2017).

Large-scale celebrations require a lot of seasonal workers. Idol-makers hire seasonal workers with a varying array of specialised skills to perform different tasks of idol-making. The tasks of the seasonal migrants include starting from the basics, like carrying bamboo slats; cutting them; and mixing the clay; to the highly skilled jobs like sculpting faces, fingers, and toes; spray painting the idol; and so on. These seasonal workers with specific skills are hired during different phases of the idol-making process. For instance an artisan skilled in clay mixing and sculpting would be hired towards the beginning of the process, whereas a different artisan with painting skills would be hired at a later phase of the idol-crafting process (interview: Kumartuli 2018).

Kumartuli is abuzz during the festival season. From the late monsoon to the beginning of the festival in autumn, the migrant population in Kumartuli is huge. People are employed not only in idol-making but also in the allied industries. Another large section of seasonal works is porting. *Kumars*, as well as the clients, employ porters. Makeshift tea stalls and food joints are also set up to cater to the large floating population. The considerable bulk of seasonal workers come from the adjoining rural areas of Kolkata for a few months of the year (interview: Kumartuli 2017). The jewellery making and the *pandal* decorating not only are confined to the previously specialising craftspeople but have also overflown to the studios or the workshops of the famous *kumars*. For the purpose of *pandal* installations and idol-making, the artists hire a large number of seasonal helps through specialised sub-contractors on a lump sum basis (interview: Kumartuli 2017). The associated industry that presents the ephemerality of festive urbanism in the city of Kolkata involves multiple agencies and presents a growing array of innovations in infrastructure and services (Chattopadhyay 2019). The list includes the construction of *pandals* and idol-crafting and ritualistic aspects of the crowd management measures put in place during the festive period. The idol is a commodity that showcases the culture[11] of a community, and the Durga Puja itself possibly pulls in a higher number of the low-paid workforce than any other informal sector industry in Bengal on a seasonal basis.

Notes

1 Referred in Table A.2, and these corresponding photographs have been numbered serially.
2 Sutanuti, Kalikata, and Gobindapur were the three villages bought by Job Charnock, who started business from the river trading hub, and later these were the founding villages of modern Kolkata (then Calcutta).
3 *Para* is a Bengali word, the meaning of which is complex and fuzzy, and rooted in the socio-spatial structures of a neighbourhood. *Para* sometimes is a place-based identity but can mean something more than just a neighbourhood or locality.
4 *Bonedi* or '*buniyadi*' means 'of the foundation'. *Bari* means a house or home. Some *Babu* families were the founding families of the urban settlement of Kolkata in the early 18th century. Most such houses were built as mansions that portrayed the owner's extravagant taste and competed among peers in terms of size and finish.
5 *Tubri* is a firework displayed during Durga Puja and Diwali in India. It is a spherical hollow structure stuffed with charcoal and gunpowder. It is made to sit on the ground and when lighted produces colourful sparkles of fire rising to great heights. *Tubri* display competitions are still held in Bengal during Kali puja festivals.
6 Uron-tubri, now widely banned for fire-safety, was a flying variant of tubri.

7 Subaltern studies represent collective groups of socially backward and underprivileged. From the classical definition by Gramsci, subaltern studies have evolved through contributions by Guha and Spivak (Spivak, G.C. (1988). Can the subaltern speak? *Marxism and the Interpretation of Culture*. C. Nelson and L. Grossberg. Urbana, University of Illinois Press: 271–313.; Guha, R. (1989). *Subaltern Studies VI: Writings on South Asian History*. Delhi, Oxford University Press.) to refer to underprivileged groups, socially differentiated by power structures and lacking recognition and representation.

8 According to customary practices, a little clump of mud from the threshold of the prostitutes' area, Sonagachhi, which is adjacent to Kumartuli, is mixed thoroughly with the clay from the river Ganges (Hooghly).

9 Durga Puja requires the sculpting of five separate idols. During the preparation phase, idols are prepared in batches, and storage in a covered and dry space is necessary until they are delivered to the clients.

10 In the predominantly Indian numbering place value system and units of measurement, 1 crore is equal to 10 million, usually used to measure rupees and people. It is the highest order in traditional system; one crore rupee is equivalent to approximately hundred thousand pounds in 2016.

11 Culture and creativity have been identified as tools that transversally contribute to the promotion of UN's post-2015 sustainable development goals (SDGs) around economic, social, and environmental sustainability. All three dimensions of sustainability in this agenda contribute to the safeguarding of culture, heritage, and creativity. These legal instruments aim at the preservation and protection of cultural heritage and the recognition of the importance of culture in development policies. While the inclusion of culture in the UN's SDGs is an exciting and innovative tool to promote UNESCO's cultural and heritage agendas, culture being broad and abstract in character brings policy challenges, more so, because most national and regional governments are driven by differentiated political will and resources. While the link between culture and development is not well-established, regional, national, and international organisations continue to place culture and heritage in policy implementation. In the 2017 UNESCO report submitted by India, it was highlighted that the Ministry of Culture is working with specific traditional art forms and social practices and rituals to preserve them. For more details, see Hosagrahar, J. (2017). *Culture as the Heart of SDG, The Unesco Courier*. Available at: https://en.unesco.org/courier/april-june-2017/culture-heart-sdgs (Accessed: 25 June 2019) and Vlassis, A. (2015). Culture in the post-2015 development agenda: The anatomy of an international mobilisation. *Third World Quarterly*. **36**(9): 1649–1662.

References

Agnihotri, A. (2017). *Kolkatar Pratima Shilpira*. Kolkata, Ananda Publishers Private Limited.

Archer, J. (2000). Paras, palaces, pathogens: Frameworks for the growth of Calcutta, 1800–1850. *City & Society*. **12**(1): 19–54.

Basu, S., I. Bose and S. Ghosh (2013). Lessons in risk management, resource allocation, operations planning, and stakeholder engagement: The case of the Kolkata Police Force and Durga Puja. *Decision*. **40**: 249–266.

Bean, S.S. (2011). The unfired clay sculpture of Bengal in the artscape of modern South Asia. *A Companion to Asian Art and Architecture*. Malden, MA, Wiley Online Library: 604–628.

Bhattacharya, T. (2007). Tracking the goddess: Religion, community, and identity in the Durga Puja ceremonies of nineteenth-century Calcutta. *Journal of Asian Studies*, Cambridge University Press. **66**: 919–962.

Bose, N.K. (1958). Some aspects of caste in Bengal. *American Folklore Society*. **71**: 397–412.

Chakrabarti, D. (2022). *Everyday Negotiations With Infrastructure and Municipal Services in Kolkata's Inner-City Kumartuli Neighbourhood*. RC21 Conference 2022: Ordinary Cities in Exceptional Times.

Chakravarti, S. (1985). *Krishnanagarer Mritshilpa o Mritshilpi Samaj*. Kolkata, K. P. Bagchi and Co. for The Centre for Studies in Social Sciences.

Chaliha, J. and B. Gupta (1990). Durga Puja in Calcutta. *Calcutta the Living City: Vol II*. S. Chaudhuri. Oxford, Oxford University Press. **2**: 331–336.

Chatterjee, J. and R. Khullar (2004). *Kolkata: The Dream City*. Kolkata, UBS Publishers' Distributors.

Chattopadhyay, S. (2012). *Unlearning the City: Infrastructure in a New Optical Field*. Minnesota, University of Minnesota Press.

Chattopadhyay, S. (2019). Ephemeral architecture: Toward radical contingency. *The Routledge Companion to Critical Approaches to Contemporary Architecture*. S. Chattopadhyay and J. White. London and New York, The Routledge: 138–159.

Das, M. (2016). Kolkata takes Rio steps with Mamata Banerjee's Durga carnival. *The Economic Times*. Kolkata.

Ghosh, D. (2012). *Pashchimbanger Mritshilpo*. Kolkata, Ministry of Information & Cultural Affairs, Government of West Bengal.

Guha, R. (1989). *Subaltern Studies VI: Writings on South Asian History*. Delhi, Oxford University Press.

Guha-Thakurta, T. (2015). *In the Name of the Goddess: The Durga Pujas of contemporary Kolkata*. New Delhi, Primus Books.

Guha-Thakurta, T. (2017). Durga, didi and the new age puja. *The Hindu*.

Heierstad, G. (2017). *Caste, Entrepreneurship and the Illusions of Tradition: Branding the Potters of Kolkata*. London, Anthem Press.

Indo-Asian News Service (2014). Durga Puja revellers line up to see diamond-studded idols. *NDTV*.

Outlook Web Desk (2016). Danger of deification. *Outlook India*.

Roy, A. (2003). *City Requiem, Calcutta: Gender and the Politics of Poverty*. Minneapolis, MN and London, University of Minnesota Press. **10**.

Roy, D. (2012). Caste and power: An ethnography in West Bengal, India. *Modern Asian Studies*. **46**(4): 947–974.

Sen, M. (2015). Craft, identity, hierarchy: The Kumbhakars of Bengal. *The Politics of Caste in West Bengal*. New Delhi, India, Routledge: 216–239, 232–255.

Sen, M. and K.B. Nielsen (2022). Gods in the public sphere: Political deification in South Asia. *Religion*. **52**(4): 497–512.

Shove, E., M. Pantzar and M. Watson (2012). *The Dynamics of Social Practice: Everyday Life and How It Changes*. London and New York, SAGE Publications: 1–19.

Sinha, S. and R. Bhattacharya (1969). Bhadralok and Chhotolok in a rural area of West Bengal. *Sociological Bulletin*. **18**(1): 50–66.

Spivak, G.C. (1988). Can the subaltern speak? *Marxism and the Interpretation of Culture*. C. Nelson and L. Grossberg. Urbana, University of Illinois Press: 271–313.

3   The spaces of production

The neighbourhood

Kumartuli is situated in the northernmost part of inner Kolkata, between the north–south spine of Jatindra Mohan Avenue (popularly, Central Avenue) and the river Hooghly. The area lies within wards 8 and 9 of Kolkata Municipal Corporation (KMC). The four roads bordering Kumartuli are Rabindra Sarani on the east, Durga Charan Banerjee Street to the north, Strand Bank Road (riverbank) on the west, and Abhay Mitra Street to the south. The eight-foot-wide road parallel to Durga Charan Banerjee Street is Kumartuli Street. A street of similar width, known as the Banamali Sarkar Street, runs almost perpendicularly to the Rabindra Sarani for the entire neighbourhood width. Most vehicular traffic is restricted to these streets, and the narrow inner lanes are usually pedestrian-only due to the width. Bicycles and motorbikes owned by local residents ply through the inner streets during the off seasons (Figures 3.1 and 3.2). The inner alleys and lanes are four to eight feet wide with workshops on both sides; during the preparation seasons from July to October, the awnings and overhangs of the workshops made of polythene sheets encroach upon the paths leaving around two- and three-foot space for people to pass through (Figures 3.1 and 3.2).

The roads are blacktopped and concrete from the wider to the narrower, respectively. The conditions of the roads are imperative to pull and distributing the idols smoothly. There are drains with broken covers and potholes in the alleys, adding to the already dingy and crowded conditions of the roads. According to the *kumars*, public works are rare and occur only before the elections. However, there were no roadworks that happened in the recent past (interview: Kumartuli 2018). One of the interviewees mentioned that residents collectively paved a small alleyway between their houses and a wider street to avoid disruption during idol distribution. Only one lane was paved in the last seven to eight years; it is now in disrepair again (interview: Kumartuli 2018). Interviewees also complained of potholes which make manoeuvring of idols during the peak-season distribution phase extremely difficult and risky.

From my fieldnotes, maps, and drawings, I narrate the (failing) physical and built infrastructure of Kumartuli and a social–cultural network that has facilitated the idol-making industry. This ethnographic account has been key to drawing the

DOI: 10.4324/9781003341222-3

Figure 3.1 A series of snapshots of the neighbourhood taken during the beginning of the preparation season in April 2018. These pictures illustrate the condition of streets, buildings, and the overall built character on and around Banamali Sarkar Street and Gopeshwar Pal Street

Figure 3.2 Street scenes of the neighbourhood taken during the preparation season in April 2018: Gopeshwar Pal Street, Durga Charan Banerjee Street, and Gokul Mitra Lane

place-based networks and processes related to idol-making practices and the mate-riality of idol-crafting. Further discussions on the built character and the utilities of the neighbourhood resources that have continued to operate despite growing pressures of the increased production function open up questions on whether the embedded multi-layered networks within and beyond the neighbourhood required a redevelopment plan aimed at formalising to function better. Based on the exist-ing physical infrastructure, built environment, and the material practices of idol-making in Kumartuli, I question the feasibility and practicability of the KMDA plan and the underpinning governmental vision to remodel and repackage the neighbourhood with aspirations of legalising the creative and production functions of the informal sector.

On mapping the neighbourhood's densely built footprint (Figure 3.3), a pattern of old and new, inner and outer, emerges from the layout. The nomenclature of the lanes and alleys suggests a pattern of emergence of the neighbourhood over time. The innermost alleys are dense and consist of rows of workshop-cum residences lined along the four-foot-wide alleyways. The core areas along Kumartuli Street and *Majhergoli* (literally, middle alley) are predominantly older settlers and potters from origins in Krishnanagar. These inner streets do not have particular names but are locally known as *Majhergoli*, parallel to Kumartuli Street, and the outer area *Bangalpara*, closer to Strand Bank Road. These areas are a little less densely built

Figure 3.3 Figure-ground of the densely built inner-core area of Kumartuli *basti*. During the peak production season, temporary sheds and awnings are built on the streets and open spaces to accommodate people and idols in the central blocks of the neighbourhood

and have larger workshops. These areas predominantly house families of Bangladeshi origin. These workshops have residences detached from them but are more organically built, mostly need-based. The outer areas have pockets of ancillary businesses. There are also a few shops dealing in ritual items and jewellery on Kumartuli Street, and these are the oldest of all such related businesses (Figures 3.1 and 3.2) (interview: Kumartuli 2018). This pattern suggests the intertwined and overlapping coexistence of idol-making and related businesses in the Kumartuli area. There are two cooperatives within the neighbourhood, with the distinction being the potters' native place of origin. The older cooperative (*Kumartuli Mritshilpi Sanskriti Samiti*) is run by *kumars* migrating from Krishnanagar, and their working office is based at the junction of Kumartuli Street and Gopeshwar Pal Street. The newer (*Kumartuli Mritshilpi Samiti*), and run by a predominantly Bangladeshi migrant potters' group, is based at the end of Banamali Sarkar Street, which might suggest a spatial divide in the places of origin (Figure 3.1).

Kumartuli *basti* is dotted with many temples housing different Hindu Gods and Goddesses. These temples, built by local landlords, different groups of migrants, and local businessmen over different times of the neighbourhood's history, bring unique characters to the streetscape of Kumartuli. This also reflects the emergence, power relations, and arrival time to the neighbourhood. While the older temples built by the local gentry are larger, outside, and accessed by more people, the temples built by the migrants are smaller and related to places associated with their ancestry, memories, and sentiments. Even the religiously significant *Dhakeswari* temple to house the deity was built frugally after the partition of India. Not only do temples carry historical, religious, and cultural associations of the community, but also they present profound place attachments and identities interwoven into the communities' everyday lives. Scholarly research in the fields of sociology, human geography, urban planning, philosophy, and environmental psychology has investigated unique emotional bonds that people assert with their surroundings, particularly places of living (Tuan 1977; Cresswell 2004; Lewicka 2010). In Bengali, the neighbourhoods are often called *para* that transcends beyond the word's literal translation and generally invokes deeper meanings and associations for a person. *Para* sometimes traverses beyond spatial boundaries and might be expressed only to associate a person's locality and attachment (Chattopadhyay 2012). In Kumartuli, residents associate the neighbourhood in part as their *para*, sometimes *kumarpara*, and identify it with the caste-based profession of idol-crafting and associated practices.

In an initiative of major slum improvement schemes to cater to general inaccessibility to toilets and sanitation facilities, the local government has built some toilets (mostly open urinals and a larger public toilet block) and drinking water taps (bottom left: Figure 3.1) in the Kumartuli area. Due to the influx of the large seasonal workforce in Kumartuli during the festive season, there are numerous public (male urinals mainly) toilets on street corners (top right: Figure 3.1). These are mainly open and inadequately supplied with water and drainage, causing odour. There is one larger public bath and toilet facility on the neighbourhood's periphery on Durga Charan Banerjee Street. The drinking water taps have been installed in the recent past on Kumartuli Street. The numbers are few, and the services are inadequate during the busier times (interview: Kumartuli 2018, 2022).

The first phase of my fieldwork ended in early December 2017. The reason being, first, because of a break in the seasonal production pattern during that time of the year, and second, a dengue outbreak in Kolkata that affected most of the inner wards of the city. A number of my respondents persuaded me not to visit them during that time, and those who did put mosquito repellent fumes in their workshops while I interviewed them. While in Kumartuli, I noticed that due to the seasonal break, workshops seemed less crowded, and the majority were being cleaned to clear out the chances of mosquito colonies. All the waste from the workshops was being burnt as mosquito repellents.

Kumartuli seemed quiet during this time. The dengue outbreak was not restricted to Kumartuli but was a citywide epidemic. One of my respondents explained residents' vulnerability to infection due to the dense population and their material practice (interview: Kumartuli 2017). Residents insisted that the rapid spread of this disease is a result of the hygiene practices of the inner-city residents (interview: Kumartuli 2017). The city government's lack of services pressurises the limited garbage disposal spaces and worsens the situation in this densely populated inner-city neighbourhood where the waste generation is more than in an average residential area. Following on from my fieldwork experiences during the dengue epidemic, I later interviewed a couple of my respondents regarding the COVID-19 pandemic in 2020–2021. The implications of the COVID-19 pandemic were severe. Not only did the densely built nature of the neighbourhood threaten community outbreak and spread, but the inadequate sanitation infrastructure also aggravated the risk. However, socio-political structures and care and moral response perceptions existing within the civil societies in *paras* helped form bottom-up networks. These networks of volunteers from *paras* supported the communities through caring, provisioning, and sometimes funding health care and related services, a phenomenon well documented in social and news media (The Economist 2021; News Desk 2022).

During my December 2017 visits, I noticed some of the workshops were being repaired and stocked up for the coming seasons. The structure of the workshops and the underlying interwoven everyday practices of family lives seemed to be visible for the first time to me. Although most workshops were empty, families continued to live and work from their residences in the inner areas of the workshop-residence units. The peripheral areas of the neighbourhood where independent workshops are situated pursue other sculpting practices than the religious idols during this time of the year. In order to study the overlapping spatial practices and patterns of everyday life within the workshop-residences and the larger workshops, I listened to the advice of local residents to study these units during the early preparation phases in late April and May. Undoubtedly, the main idol-making practices in Kumartuli are still dominated by Durga Puja in early autumn. Traditionally, idol-making started in the monsoon after the festival of Rathayatra. However, the growing demand for idols globally and the increasing spectacle of the festival are slowly altering the seasonal practices in Kumartuli. The new seasonal pattern of idol-making practices starts in late April and May (*Boishakh*, the first and auspicious month in the Bengali calendar), followed by the busiest monsoon work span. The winter months of December and January (*Poush*, an inauspicious month) are the slowest in terms of business (interview: Kumartuli 2017).

Although idol-making and allied practices are generally governed by the contemporary tourism-driven economy surrounding the Durga Puja in Kolkata and beyond, interestingly, the idol-makers and their clients still follow some traditional rituals, beliefs, and religious auspices. While a group of *kumars* craft smaller idols for household consumption following these traditional norms, larger and higher budget idols are produced without much consideration for rituals and seasonality, some functions of which have started to move out of Kumartuli. However, even the busiest workshops only build the *kathamo* (wooden frameworks) during March and early April, carrying on repair works while keeping the rest of the production for later auspicious months. Factors like the availability of seasonal help or lack of funds to start preparation might also be responsible for such a pattern of seasonal practices, suggesting that the current operations of the industry in Kumartuli are still predominantly seasonal. Unless idols are commissioned early or they have to be consigned abroad, the production functions are still mostly seasonal.

Streetscapes, alleys, riverfront

Although the Kumartuli *basti* is on the riverfront, it is not flood-prone. Even during the highest tides of the river and heaviest monsoons, Kumartuli does not face the risk of waterlogging, which is quite significant for clay-idol-making practices. The two reasons for low flood-risks are that the river Ganges has certain natural levees and high embankments built during the colonial period, and a central city location provides links to the municipal drainage system. The riverfront has positively contributed to effective idol-making practices in Kumartuli by providing clay, transport linkages, and associated rituals of idol-making. The *ghats* (riverfront embankments) are extensively used to source and store items like bamboo and clay during the preparation phases, and these spaces are associated with the transaction of the raw materials from the suppliers to the *kumars* (Figure 3.4).

The riverfront roads are wider and well-connected to the rest of the city. The riverfront road has restricted access to public transport; only smaller vehicles and goods trucks are allowed. Although it might not be the only reason, the width of the roads and the allied infrastructure have positively contributed to the continuation of rituals and material practices related to idol-making in Kumartuli. Also, the network connectivity of the inner-city location and ease of access to physical infrastructure have led to the promotion of Kumartuli as a cultural industry by the local government. However, the value of the inner-city real estate, associated rent, and continued promotion of the idol-making industry have led to the displacement and gentrification for some residential and productive functions elsewhere. Implicitly, there is a tremendous drive within Kumartuli to move out organically and retain

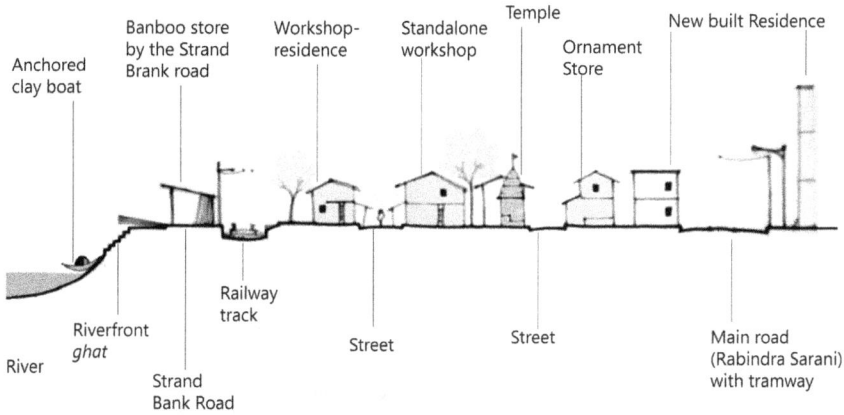

Figure 3.4 Schematic section through the *ghat* and riverfront road

Source: author

storage spaces elsewhere in the city. This poses questions about the storage and redistribution of finished idols.

The average day in a workshop-cum-residence is an overlap of the practices related to the production of idols and everyday practices of living in a densely built *basti* neighbourhood of Kolkata. The streets in front of each workshop are extended space that serves many purposes: it is essentially an integral part of the everyday activities of the residents and sometimes acts as an extension of the workshop itself. Curious visitors, like tourists, frequent these streets, and photographers and researchers allow the residents little or no privacy. Both sides of the internal alleys are lined with workshops; some have attached residential units at the back accessed through the workshops, while some residences have separate entrances through narrower alleyways, just enough to allow one person to walk through. Additionally, the streets are used for an array of activities related to idol-making like clay-sorting, mixing, bamboo and straw cutting, and even sun-drying of idols or parts of them after clay-application (Figures 3.1 and 3.2). The largest of the workshops are at street crossings or on the wider roads.

Kumartuli is currently undergoing a transition from being a fairly mixed-use towards becoming an idol-producing, significantly more commercial neighbourhood. A wave of mostly organic gentrification has resulted in the changes to the primary use of the buildings in Kumartuli (Figure 3.5). Several older structures have been demolished and rebuilt to suit the purpose of workshop spaces. Some dwelling spaces have been rebuilt into multi-storied multiple family homes behind the workshops. Some families have outgrown the spaces, some have been able to afford better living conditions elsewhere, and some have been rehoused in the nearby warehouse, modified to suit the purpose of the temporary needs of the Kumartuli rehabilitation project of the state government and KMDA. What used to

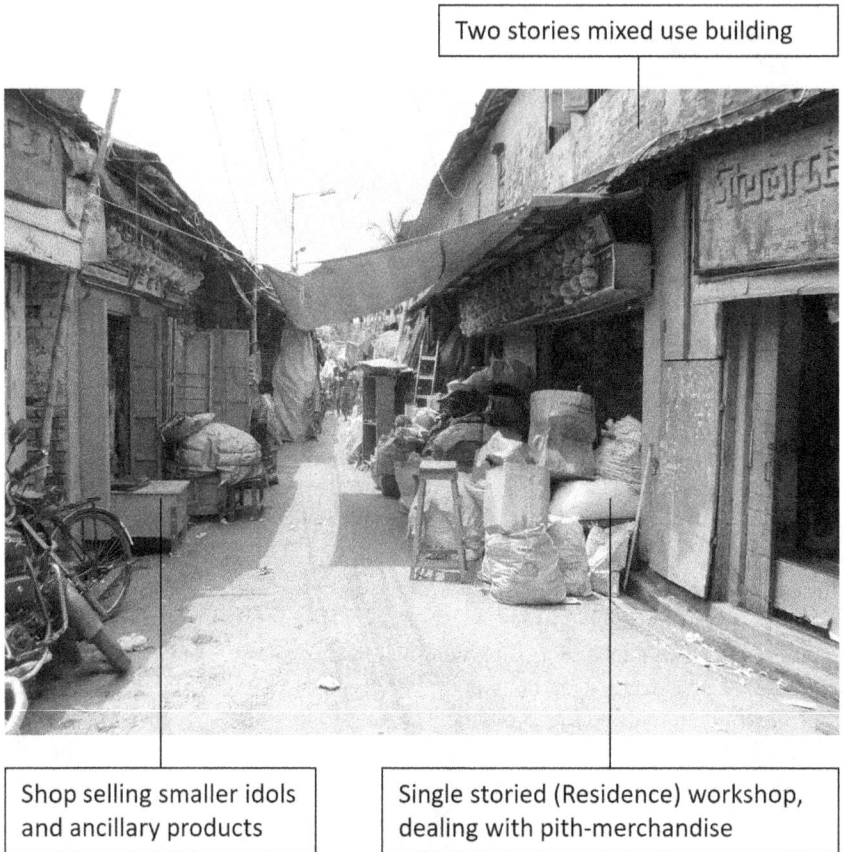

Two stories mixed use building

Shop selling smaller idols and ancillary products

Single storied (Residence) workshop, dealing with pith-merchandise

Figure 3.5 A predominantly mixed-use street: *Majhergoli* of Kumartuli, photographed in early May 2018

Source: author

be a caste-homogeneous neighbourhood of potters' families living and working in cramped conditions is now transforming into an area of heterogeneous, primarily productive functions driven by tourism and the cultural economy.

Three distinct types of spatial configuration were identified on the basis of my observation and subsequent spatio-temporal study on the layout of the workshop-residences of Kumartuli and the practices therein. The first type was the earliest flexible workshop-residence built for the use of a single family of idol-makers. Quite a few of this type of layout still survive in the inner alleys of Kumartuli. The second workshop type looks like a 'factory shed', purpose-built to accommodate idols (and seasonal workers). These are situated around the outer streets of the neighbourhood and are comparatively larger in area. The converted or transitioning houses are a third and possibly the more common type of contemporary workshop.

These serve the purpose of a shop, workshop, and accommodation, only used during working hours. Currently, in Kumartuli, a fourth type of workshop is emerging due to the recent confusion and ultimate failure of the KMDA redevelopment project. This type represents the new-built or temporary sheds, constructed after the demolition of a few workshops by KMDA and solely to retain a workspace in Kumartuli. Recent socio-political conditions, perception of quasi-legal opportunities, and years of governmental inaction have resulted in rampant demolition and rebuilding works within the neighbourhood.

The government policies on displaced communities in Delhi, India, as observed by Routray (Routray 2014), pose questions on perceived notions on views on legal, social, and political systems in India (Chatterjee 2004; Chatterjee 2011). The views of politicians, state officials, and civil servants elaborated on ideas of enforcement of norms and questioned the ideas of culturally deprived communities in the informal urban settlements. Informed by Routray's arguments about political society and modern governmental systems in India, a contextual understanding of marginal communities and their lack of participation and resistance to governmental reforms can be elaborated on. As a citywide understanding of neoliberal urban planning reforms in India is imperative here, this study provides scope for analysing the inequalities and ambiguities of legal enforcements based on '*hierarchies of caste, class, communities and gender*' (Routray 2014).

The conventional workshop-residence

The modest home of the potters (*kumar-bari*) served as the hub of idol-crafting in the past. Within the cramped confines of the idol-makers' quarters in Kolkata, the potters managed their daily family lives while also crafting pots and clay idols. Over the years, idol-making practices have changed. Also, the spaces used for performing these practices have adapted, grown, and evolved. The workshops house the idols being built, raw materials, tools, and all the necessities needed to provide support and house the idol-maker and his assistants while they are working. Also, most workshops have a *macha* or mezzanine level made of wooden slats to provide for the lodging of seasonal workers and storing finished idols. This arrangement perhaps is important to accommodate the seasonal workers in the workshop during the day to avoid delays in work. Very simple and affordable construction methods have been historically used to construct the often self-built houses.

The design principles are always need-based. Most houses in this neighbourhood have brick walls plastered with cement concrete and roofs of temporary materials. The roofs resemble 'factory sheds' made from galvanised steel, tin, asbestos, or PVC sheets (interview: Kumartuli 2018). The houses are between 18 and 20 feet high, with internally constructed mezzanine levels made from wooden or bamboo slats extending from the front to the rear of the houses. A ladder is generally used to access the *macha* to save floor space and the additional cost of staircase construction. The street-fronts and open areas are covered with plastic sheets, temporary bamboo sheds, and awnings during the prolonged monsoon season to assist the smooth running of idol-making and other related everyday practices in Kumartuli

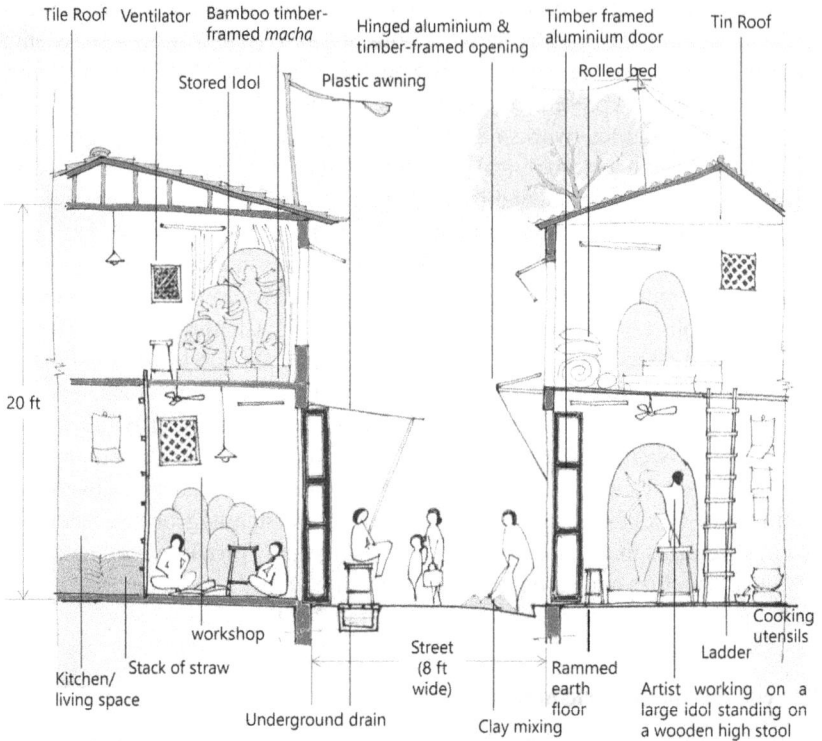

Tile Roof   Ventilator   Bamboo timber-   Hinged aluminium &   Timber framed   Tin Roof
                         framed *macha*   timber-framed opening   aluminium door

Stored Idol   Plastic awning   Rolled bed

20 ft

                workshop   Street   Cooking utensils
                                            (8 ft   Ladder
Kitchen/   Stack of straw   wide)   Rammed
living space   earth   Artist working on a
              Underground drain   floor   large idol standing on
                                            Clay mixing   a wooden high stool

Figure 3.6 Section through a typical street in Kumartuli, showing workshop-residences and
            the sociocultural spaces

(Figure 3.6). Therefore, during this time, the workshops encroach upon considerable space from the streets and adjacent common areas to accommodate all idol-
making activities.

A typical workshop-cum residence has the front room, which is usually a third
of the entire length of the house, dedicated to idol-making and daytime activities
of the residents. This front room has foldable wooden-framed tin shutters along the
entire front wall, which can be opened completely to allow light, ventilation, and
movement of finished idols. The back rooms are usually for the family to perform
the everyday practices of cooking, cleaning, and sleeping. This part of the house
has one bedroom, kitchen, and dining space. A few families share separate toilets
and washing facilities at the back of the house. This style of workshop-residences
is situated predominantly on both sides of the *Majhergoli*.

The first of this type of workshop-residence I studied in detail was the residence
of two women, who were sisters-in-law. One of the women had continued to practice her husband's profession after his demise. At the time of my fieldwork (in the
summer of 2018), she advised a group of seasonal workers who continued carrying
out a major share of labour for idol-making with her through the peak seasons of

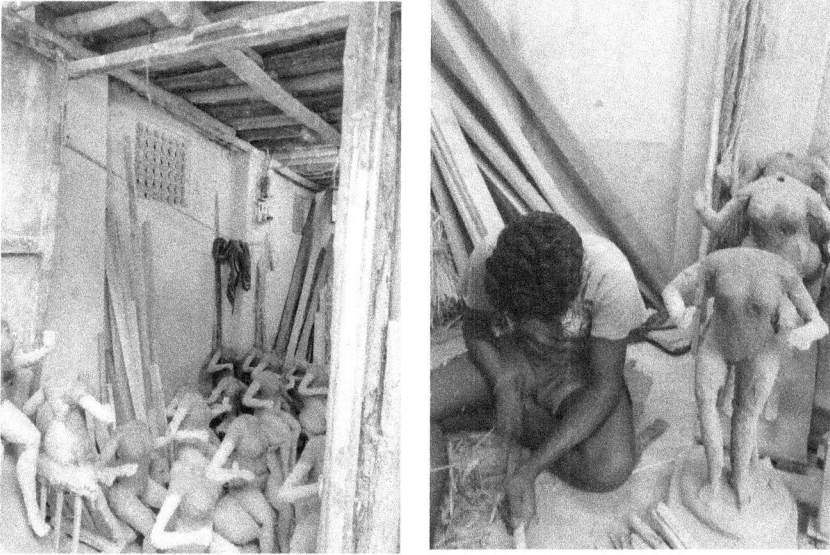

Figure 3.7 Workshop of the lady, with idols stored, and the younger seasonal migrant employee at work

the last ten years (interview: Kumartuli 2018). During my study, two men were working in the front room of the house (Figure 3.7). The front room has a mezzanine floor with a low ceiling height of about eight feet which is accessed by a ladder made of bamboo and wood. The upper storey is used to keep smaller idols, decorative materials, and outfits for the idols. During the peak-production phase, this space is also used for lodging seasonal workers. Just across the street, around eight to ten feet away, is the workshop-residence of an elderly artist. This workshop, facing the former, is a couple of feet higher and wider. However, the room layout is similar, with a front room dedicated entirely to idol-making practices and the rear part for the family of five. In this workshop, the older artist was working alone on the day of the study, in early April 2018.

The floors of both buildings are made of cement concrete. Throughout the months of production, clay is continually used and stored with raw ingredients, giving the flooring a damp and earthy appearance. Interviews have revealed varying opinions regarding the kind of flooring most suitable for crafting idols. Given their limited resources, some more experienced artists contend that the floor should be as level and hard as possible (interview: Kumartuli 2018). However, famous contemporary artists have suggested that they require soft earthen floors; soft floors are useful to avoid cracks and shattering of freshly sculpted parts of idols with soft clay on accidentally tumbling (interview: Kumartuli 2018). Although this explanation is offered for the idols' fingers and faces being delicate during preparation, the majority of the floors I have seen were levelled, reasonably flat, and constructed of cement concrete that was set in place. Most workplaces have repairs done in early

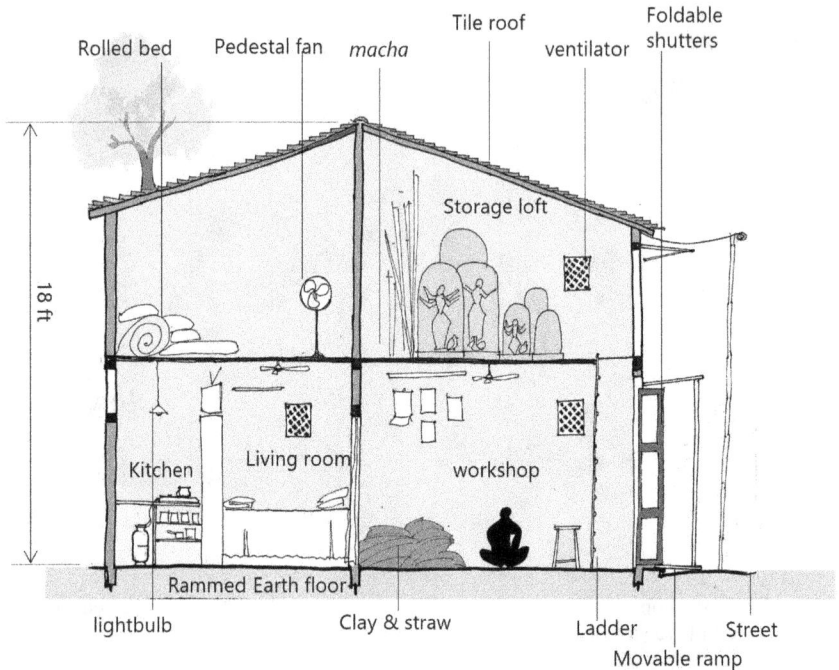

Figure 3.8 A section of a typical workshop-residence

April, just before the annual preparation phase begins. The damaged floors of a few workshops I came across were being patched to have an even surface; otherwise, *'the faces of the idols would be tilted'* (interview: Kumartuli 2018).

The walls of both buildings support the roof and the bamboo-framed *macha* (see Figure 3.8). Most of the houses constructed in poorer neighbourhoods in Kolkata use a combination of vernacular and contemporary building methods, with the walls being coated with brick and cement mortar and a pitched tile or tin roof. A system of wooden columns based on the foundation supports the roofs. According to the Census of India (Ministry of Home Affairs 2011), these buildings belong to the semi-pukka category because they are constructed of *kutcha*, transient roofs, and *pukka*, or permanent walls. The structural duality and materiality impact the appropriate property taxation criteria. The annual property taxes are determined by considering several important criteria, including the width of the road, the kind and age of the buildings, the building materials and nature of the built structure (*kutcha/pukka*), and the location of the property (KMC 2017). Due to contributing variables like narrow roads, dated and blighted structures, semi-*pukka* dwellings, and a registered slum location, residents of impoverished *basti* neighbourhoods benefit from lower property taxes.

Views are divided about *pukka* and *kutcha* nature of the walls. Sunlight and steady airflow being two of the foremost requirements to avoid damped conditions, and the artists are all of the opinions that walls should be able to provide ventilation (interviews: Kumartuli 2017, 2018). However, the only openings are in the front and back of the houses, as most of the sidewalls are shared with neighbouring structures. There are some grilled ventilators (locally known as *jail*) in between the walls of shared houses. The roof of the smaller house has a wooden and bamboo-framed pitched structure and is covered with terracotta tiles. The larger building across the road has a tin surface above the timber frame structure.

I studied the use of the spaces and movements within these two workshops over a period of six hours (see Figure 3.9): three hours in the morning starting from 9.30 am and three hours after the lunch break starting around 2.30 in the afternoon. The methodological details of this study and the exemplar of the corresponding drawings produced during the fieldwork are presented in the methodological appendix (Figure A.1). This study involved mapping the activities and movements within the house and the street in front. Moreover, I mapped only activities related to idol-making and not everyday activities like cooking, eating, sleeping, and using toilets.

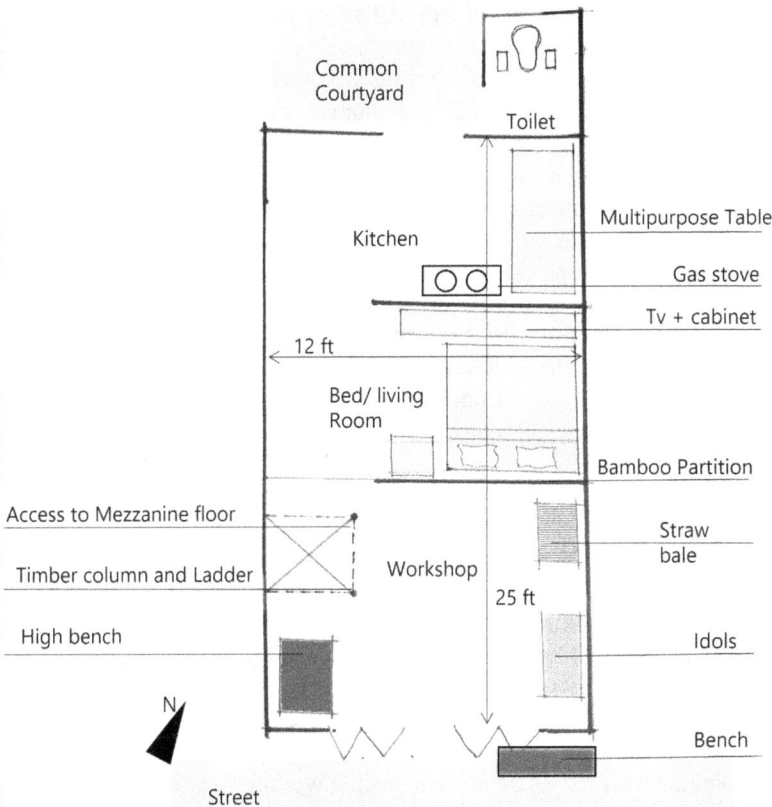

Figure 3.9 Details of a typical workshop-residence

However, breaks in the idol-making pursuit due to these activities were marked to record the idol-making practices in particular and how these are interwoven with the everyday activities of Kumartuli residents. This mapping helped to understand how much space is required within a workshop to perform the essential activities related to idol-making, and how the interior space is arranged accordingly, as most of the workshop-residences do not have clearly delineated boundaries or, in particular, space for the specific stages of idol-making. While practitioners actively work within cramped yet flexible spaces (Lefebvre 1991), these continue to be shaped by the materiality and the practices performed within them.

The workshop walls are modestly decorated with pictures of idols previously crafted by the artists. At one corner, there was a small box hanging on the wall containing very small idols (around six inches) of *Lakshmi* and *Ganesh*, worshipped on the first day of the Bengali month of *Boishakh* (mid-April) every year, which marks the ceremonial beginning of the business year. Idol-making practices are rooted in Hindu religious customs, where idol-makers are essentially crafting Hindu gods. Hence, rituals customary in Hindu religion and culture shape the meanings and beliefs within their everyday practices. Every morning, before the start of the day's work, the artists show respect for their beliefs by cleaning the workshops and traversing all corners of the workshop with lit incense sticks. For a few minutes after this daily ritual, the workshops smell pleasant, overpowering the nearby odours of carelessly disposed of garbage and other human and animal wastes on the street. During the day, the activities are concentrated in the front rooms of the house. The *malik*, or the main artist, manages the work of the seasonal workers. The front rooms have all tools and equipment required within arm's reach: chisels, hammers, saws, spades, and shovels. The only type of furniture in these rooms is varying sizes of footstools and benches, precariously built to serve the purpose only. Heaps of straw, clay, buckets, or bottles of water to drink and work with are stacked in the corner of the room before starting the day's work. The workshop receives adequate sunlight during the day through the wide shutters, which remain open throughout the working hours. A ceiling fan hangs from the *macha* above, which makes a constant humming noise. After a while, this noise almost mingles with the constant chiselling and hammering noises inside the workshop and feels like a part of the entire ambience.

The men use the floor for sitting and working on the larger sections of the body of the idol (Figure 3.7). These larger sections include sculpting the idols' hands and legs out of straw. Bunches of straw are tightly bound with jute or coconut fibre-twined ropes to shape individual parts of arms and legs. This aspect of idol-making requires strength and effort to hold the straw and tightly bind it to form the shapes of cylindrical arms and legs. Hence, this job is usually carried out by the younger seasonal workers. Across the street, the older artist was shaping idols' fingers and faces out of soft clay (Figure 3.10). His workbench was placed on the outermost part of the workshop at the threshold of the street. Here, he was able to receive enough light and air to feel comfortable while working. The shutter of the *macha* was opened up and supported on bamboo from the street. This arrangement would provide ample shade during mid-day and the afternoon. He used the same workbenches for sitting as well as his work surface.

Figure 3.10 Workshop-residence: workshop of the older artist and the street workshop spaces

Source: author (April 2018)

The tools used were simple: thin sculpting chisels and brushes. Water from a small bowl placed on the wooden bench was used to smoothen the rough edges. When finished, these finer parts were left on slabs of wood by the workshop's entrance or on the street in front to dry in the sun. The men kept a watchful eye on these during the day to avoid any damage caused by passers-by: children, stray animals, and tourists (interview: Kumartuli 2018). In a corner of his workshop were sculpted faces lined up on a wooden slab, which had been finished the day before. The floor of this workshop was clean, as work for the year was yet to begin in full swing.

I observed a striking 'seasonal' variation from these two seemingly similar workshop-residences. In the first case, the seasonal workers had already arrived; hence, it was busier with two workers and the artist. Everyday practices in the first house revolved around the front room, which was the living-cum-working space. However, the second workshop was being used by one person. The floors had no stored raw materials except a small lump of *entel-mati* (soft-sticky clay) and a couple of buckets full of water.

Around 10 am, after completing cooking and cleaning, the *malik* of the first house came to watch over her seasonal workers and advised them on the day's work. She placed her bench on the street, touching the threshold of her property. From here, she could keep an eye on both the workshop and the street. I kept recording what went on in the workshop—how the two younger men tied the knots to secure the shape of the sculpted straw, got up to fetch more straw, kept the completed pieces, resembling the shape of limbs, on one side, and moved on to the next. The younger, and perhaps the novice of these two men, always asked for approval from his senior and *malik*. After a while, he was asked to cut a new bunch of straw to the required size, which he did—he placed a wooden board close to the doorway and carefully cut it with a hacksaw. Then he cleared the floor of the offcuts with a broom and left these at the corner of the room by the street. After a while, the other man went out for a smoke and came back around ten minutes later. The younger man went to the toilet outside; a public urinal is just around the corner on the street.

Across the street, the older artist took a couple of breaks: he smoked a cigarette while sitting on his bench, stood up and stretched his legs, went inside his house behind the workshop for a toilet break and picked up the batch of clay-sculpted fingers from the workbench and placed it on the floor for drying. Just after midday, he went up to the *macha* and brought a picture album for me to see. In this album, he kept a record of most of the idols he had crafted during his long career, spanning over 35 to 40 years. Also in this album were newspaper cuttings, pictures of him in Kumartuli, his prize-winning idols, and stories of him exporting smaller and light-weight idols of Durga abroad covered in the local newspaper. I appreciated his achievements and thanked him for showing me his picture album. This album was, in fact, a portfolio that he would show to a potential client. Even after so many accolades, this 68-year-old man led a humble life in a modest house in Kumartuli and wore a cotton dhoti and a simple white cotton shirt. After the conversation, he pulled the plastic sheets on the shutters like a curtain and went inside his house for lunch and an afternoon nap.

On the other side of the street, the lady put the bench inside and went into the house for lunch at around 12.30 pm. One of the men cleaned the floor, just enough to sit down for a meal, at the centre of the room under the ceiling fan. The other man went to a nearby tea-stall canteen (on Rabindra Sarani) to buy a plain rice and fish meal. I enquired whether they buy lunch every day. He said that on days when more people are working, they cook a meal on a gas stove in the workshop, and on days like this one, they buy: '*whatever is convenient really*' (interview: Kumartuli 2018).

In the afternoon, just after lunch, the seasonal workers laid out a mat and rested until late afternoon. They started working again at 2.30 pm and continued until evening. In the late afternoon, the lady made tea for all the men, herself, and me. A neighbour, a woman who lives two doors down the street, usually comes to visit her every afternoon for a chat. They crack jokes and gossip about other neighbours and laugh. My presence somewhat shortened her visit, and she went to visit yet another neighbour down the road. I had to apologise for the inconvenience I was causing, and they reassured me that there are tourists, even foreigners and film-makers, who visit them sometimes and ask to watch them work. The older man from across the street reappeared in his workshop shortly after 3 pm. He opened the doors, removed the plastic sheets, brought his bench to the front, and sat with a cup of tea. He did not immediately start working. I asked whether or not he would reconvene his pursuit for the day. Unlike the woman across the street, he had not yet received orders for that year. Also, his seasonal employees were engaged in idol-making elsewhere across the country. Therefore, he would wait a few more weeks to start work for Durga Puja. In the meantime, he was advancing his work during his leisure. My enquiries triggered a conversation on idol orders among the neighbours, who were concerned that work would commence later than usual due to the delay in the festive season on that year's calendar. They continued talking; local children walked back from school through the alleys, some peeped inside the workshop, the owners greeted some children, and others passed by curiously, looking at me. The street was livelier in the late afternoon and early evening; more people had opened their workshop doors. Streetlights were slowly lit, and nearby shops opened up. A while later, a group of men came up to the workshop to ask for their friends, and the men went out for a smoke and decided to end their day's work. They switched on lights inside the workshops as it started to get dark. In the evening, both the *maliks* cleaned their workshops and again lit incense sticks. After recording the day's work on my map, I left. As this was not a particularly busy time of year, the work ended earlier than usual. However, during seasonal peaks and nearer to deadlines, workers work late into the evenings and even overnight.

The facing workshops and the closely packed houses provide a constant inter-action between the resident families. Both artists place their workbenches on the threshold of their workshops or the street during the day. Both share storage, accommodate each other's seasonal workers, and combat common problems by sharing resources. The seasonal migrant workers, who come from the same areas, support each other with everyday chores and emotionally. They live in accom-modations within the mezzanine floor of certain workshops; a few artists share such provisional lodging spaces. Also, during the peak production season, many seasonal workers sleep on the workshop floor between the idols. This coordina-tion exists between generations of artists and neighbours in Kumartuli. Although this highlights the social capital in the once caste-homogeneous neighbourhood, it does not necessarily suggest that there is no visible hierarchy within the spaces of production.

The practices related to idol-making being performed within the same prem-ises of their residences overlap with their private everyday lives, like household practices. The older artist mentioned that their forefathers have laid out the norms; the employer must arrange the provision of food and lodging. Only this part of the norm is stated; however, the power relations are such that the employees are restricted within the spaces of idol-making practice and using public areas to do other everyday activities. Despite these norms, interviews suggest that seasonal workers prefer to work with the same *maliks* over many years due to familiarity and vice versa. Not only do the artist and their family enjoy the privilege of using the entire building as their own, but also they govern how the spaces are organised and used. The seasonal workers, despite sometimes lodging in these workshop-residences, use the workshop space from the street threshold to the inner space of the workshop and the mezzanine floor. They restrict their entry to the bedroom and kitchen of this particular house primarily because two women inhabit it. However, this only suggests that they abide by a social norm of privacy in the inner areas of their employer's homes, an idea that has been a prerogative of the Bengali *bhadralok* (Chattopadhyay 2005).

Spatial practices drawn from this study show temporary reorganisation within the neighbourhood. Multiple relations like the arrival of migrant seasonal workers from rural Bengal, the commissioning of idols, and the transaction within the supply-chain networks trigger this change in character. The idol-making practice combines the festival's materiality, sociocultural processes, and seasonality, as well, how the production places constantly adapt to the multiple relational perspectives and the industry's temporal changes with accommodating growing numbers of people within the same spaces over time. The traditional types of workshop-residences illustrate a part of this phenomenon, which is revealed further in the following chapters.

The 'factory-shed' workshop

In late April 2018, I studied the standalone workshops on the wider streets of Kumartuli. This style of workshop is situated around the periphery of the core *Thakurpotti* (literally translating to idol-cluster) on Banamali Sarkar Street (Figure 3.3). These are not connected to the residences and are quite large compared to the front-room workshop-residences (Figures 3.11 and 3.12). The length and width of this space are almost double the size of the front-room workshop-residences. The *macha* is built higher than the smaller workshops and provides more headroom at the top. The total height of the workshops ranges from 20 to 25 feet (interview: Kumartuli 2018). However, the basic spatial organisation within the workshop is somewhat similar. The building has a rectangular plan with openings at the front and back; the front door is fitted with foldable shutters that open up completely to allow light and ventilation. There is a small window at the back. The length of the workshop is more than 40 feet; hence, the sunlight does not penetrate to illuminate the middle portion of the workshop and requires artificial lighting throughout the day. Lighting is simple—incandescent bulbs hang from holders connected through wires precariously tied to the beams. Most wirings are not concealed, providing scope for moving light fixtures around as and when required. The ceiling fans are hung from the beams. A timbre and bamboo framework, with timbre columns and beams, support the roof. The *macha* is also made of bamboo slats and plywood boards. Unlike the smaller houses, the *macha* in this building does not run the entire ceiling length; it ends about halfway through, starting at the back so that there is a higher ceiling at the front of the house to accommodate taller idols.

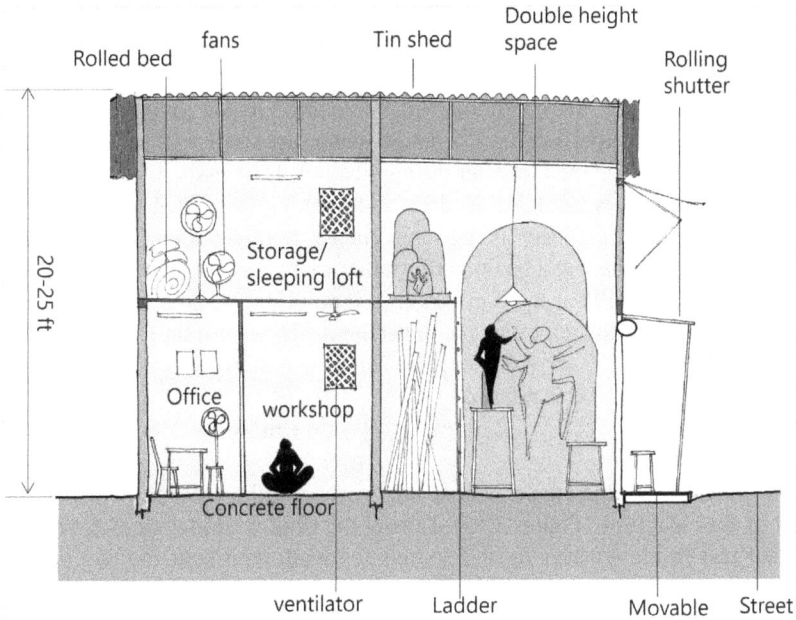

Figure 3.11 A section of a 'factory-shed' workshop

Figure 3.12 Two larger workshops showing (on the left) the internal layout during the afternoon break and bamboo-slat roof with polythene insulation, April 2018, and
(on the right) the artist applying final touches to smaller idols, October 2017

In an interview, a respondent explained with mixed emotions about their building requirements and how this precarity affects their everyday lives;

> [F]or idol-making, the condition will be damped if you give me a pukka-structure. We need openness. If it is too damp, the straw will be rotten. The clay will not dry on the idol after the application. The tin-shed, what is called a 'factory-shed', is what we need for idol-making. . . . Then the heat (and light is required). . . . Rain is our main enemy. Sunlight is our main energy. We will feel the heat and be uncomfortable; we will sweat and switch on the fans; only then will our work progress. If it rains, our work is hindered. This is the problem. Not a lot of people understand why we use tin-shed. Tin-shed is not for show; it is for wind and air. This is our workplace. . . . This is for our work; this is what is required.
>
> (Interview: Kumartuli 2017)

The workshops are the spaces where the artists and the seasonal workers spend most of their daytime (Figure 3.13). During the peak working seasons, the artists eat and rest in the workshops and sometimes work overnight to meet deadlines

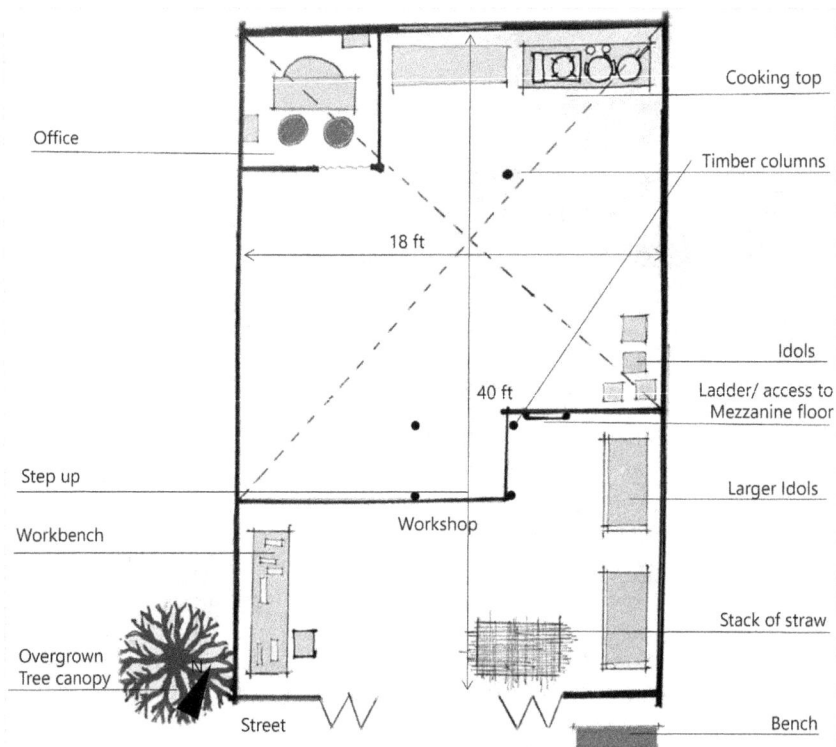

Figure 3.13 Map of 'factory-shed' workshop

(interview: Kumartuli 2018). The standalone workshops are not attached to residences, although most artists live within walking distance from their workshops, and the seasonal migrant workers are accommodated within the workshop or somewhere nearby.

Interestingly, there is a small room in the corner of this workshop. The room remains locked during a major part of the day, whenever not in use. The room contains a desk and a few chairs and usually serves as office space for the main artist. The walls of the room are decorated with pictures of idols crafted in the past, a calendar, and a small wooden shrine containing figurines of Gods regularly worshipped. A fan, an incandescent light hanging from the ceiling, and a few papers and stationery complete the list of items in the room. The wooden partition used to build the wall is a later addition and is visible from the age of the building. The office is significant not only because the idol-making practices have transformed from being a family affair based in residence to standalone workshops but also because they show the requirement of client meetings and record-keeping for the professional practice.

Like my study of the movement and spatial configurations in the earlier workshops, I conducted a six-hour mapping exercise (three hours in the late morning and three hours post-lunch). Unsurprisingly, a few routine activities are quite similar in all workshops. These everyday activities include cleaning the workshop, clearing the floor, placing the workbenches in the corners, filling the water buckets, and performing the religious ritual of lamp and incense stick lighting. However, the basic difference is in the scale of operations.

During the day of the study, four men were working in the workshop. The *malik* (main artist), a middle-aged man, worked on small clay figurines that were used as accessories for the idols. A large heap of straw was stacked on the other front corner of the room; this space was well-lit and ventilated. Next to the stack of straw, a couple of large unfinished idols stood—these were not sold last season. Therefore, the artist, facing a financial crunch at the beginning of the season, did not get a return on investment from the previous year. The pace of that year's (2018) work had been slow; yet he had to employ three seasonal workers to keep up with the preparations so that he could deliver on scheduled dates when the orders came in. Dates are important in their practices, '*the calendar does not wait for anyone*' (interview: Kumartuli 2018). The three other men were first briefed about what they were planning to do. One of the younger men was asked to fetch a bag of clay from the *mahajan* (supplier) by the river. He took the *malik's* bicycle and left. In the meantime, the *malik* phoned the *mahajan* and told him to record the transaction as credit in his ledger. After exchanging pleas and persuasive words, the *mahajan* agreed but warned that this would be the last time. The main artist explained that due to the growing competition in Kumartuli, the suppliers were no longer ready to work on verbal agreements and credits. The network of actors in Kumartuli operates on personal connections developed over generations. The growing demands and cost of raw materials have altered transaction processes, indicating a more formal and translocal transaction across wider geographies of actors.

The other two men started working on a wooden framework for an estimated idol. Because the idols had not yet been ordered, the main artist relied on his guestimates and previous experiences to prepare for that year. Towards the middle of the workshop, the two men carried on cutting the wooden slats into pieces and joining them together with nails. The sawing and hammering continued for another half an hour until the first man returned. One of the men helped him carry the bag of clay and placed it in the corner of the room by the workbench. He then took a spade and mixed part of the clay with water. Then, he placed the mixed lump on the workbench for the main artist to work with. After mixing the clay, the youngest man went and washed up from the nearby communal water tap. He came back and prepared to cook lunch for the three of them. At the inner corner of the workshop by the window is a gas stove, a few jars, and a couple of pots and pans. He continued cooking past midday until the other two men broke for lunch. The *malik* prepared to go home, which is five minutes walking distance from his workshop, while the two men went for a quick dip in the river.

When I returned after the lunch break around quarter past one, I saw a heap of uneaten rice disposed of by the drain opening for the stray animals to feed on. The men called out the stray dogs by name, and a couple of dogs hurriedly approached the food. After washing up, the men rest for a while before resuming work at around 2 pm. During this time, I interviewed them and asked about their year-round employment. Unlike other seasonal workers, one of them was a carpenter who works around Kumartuli with several main artists as a freelancer. The other man who had been helping him construct the frames (*kathamo*) was learning his trade and was employed by the first man. The youngest of the three men was also an idol-making 'intern' learning the craft from his *malik*. He had travelled from *malik's* native village in Ranaghat (near Krishnanagar) through a family kinship and resided in the workshop *macha*.

The *malik* returned shortly after 2 pm and offered to buy me and his employees tea before they resumed their work. The carpenters kept on working on their frames while the intern started helping out the *malik*. They were sculpting small figurines of animals and other accessories for the idols. The earlier batch of clay figurines was left to dry in the inner corner of the room. Over the next few hours, all the men took a couple of toilet and smoking breaks, all of which involved stepping outside the workshop. Shortly after 5 pm, the carpenters finished working for the day; they moved the assembled frames to one side of the workshop, packed their personal tools (planes, chisels, saws), and prepared to leave. I left as evening fell, and the *malik* started his evening ritual of lighting the incense sticks and candles.

Unlike the residence-workshop, the standalone 'factory-shed' workshops do not have a diurnal variation in activities. These standalone workshops are not used for living during off-peak seasons, except for a few cases like this one where one or two interns came to learn their trade. Also, these workshops are used for storage purposes during off-peak seasons—for raw materials, unsold idols and tools, extra bedding, and fans. The spaces within workshops were not meant to be hierarchical; however, locked-up office rooms and particular workbenches are usually off-limits for employees and seasonal help.

An interesting outcome of the spatial study of the 'factory-shed' workshop is that it reveals a changing phenomenon of idol-making practices shifting from a family business to a more collaborative practice. The presence of the carpenter and his employee brings into the equation a new stakeholder who apparently had not been visible since the beginning of my fieldwork. The carpenter's role is limited to preparing the idol's *kathamo* (wooden framework), which so far was created by the *kumar* himself. The increasing size of the idols, which require skilled carpentry knowledge, and growing demand have resulted in employing carpenters to complete this job. An increasing number of local carpenters are contracted for a month or so by the idol-making trades to construct as many larger *kathamo* as commissioned. While my spatial study was limited to workshops in Kumartuli, during the interview of one artist outside the neighbourhood, I saw that his permanent workforce constituted of sculptors, trainees, and carpenters. It is also important to mention here that this interviewee, *kumar*, is also contracted to sculpt contemporary art installations rather than idols only (interview: Baranagar, Kolkata 2018). This reveals increasing employment opportunities and the slowly expanding heterogeneous socio-spatial landscape the industry shapes within the wider frame of the postcolonial urban geography.

The built form of the neighbourhood, including the workshops, residences, and physical infrastructure, is evocative of abject characters associated with slum settlements in cities of the global South (Bhan 2019). Positioning this research at an intersection of Lefebvrian analysis of the production of space (Lefebvre 1991) and everyday experiences interwoven with idol-crafting practices, I study idol-crafting practices in Kumartuli. This narrative opens up questions on infrastructural entanglements (Chakrabarti 2022; Iossifova et al. 2022) and the notion of heterogeneous infrastructural configurations prevalent in geographies of the resource-poor global South (Lawhon et al. 2018). The evidence presented in this chapter reiterates eco-socio-spatial vulnerabilities and precarities that continue to marginalise the Kumartuli community. Drawing attention to the built form, and spatial configurations, I argue that while Kumartuli is a notified slum where the primary economic, productive function is thriving, residents continue to shape, adapt, and negotiate cramped and failing infrastructural system to curate their livelihoods. They shape and adapt workshop spaces within the limits offered by the planning deregulations of an informal *basti* neighbourhood, and this flexibility has resulted in heterogeneously built multiple typologies of production and commercial spaces. Research participants have iterated and reiterated the importance of governmental support, provision of basic public services, and repair and maintenance of existing infrastructure through our interactions during my fieldworks in 2017 and 2018 and during the participatory workshop in 2022 in the aftermath of COVID-19 pandemic.

References

Bhan, G. (2019). Notes on a Southern urban practice. *Environment and Urbanisation*. **31**: 639–654.

Chakrabarti, D. (2022). Transitioning infrastructures and socio-cultural practices at the idol-making cluster of Kolkata's Kumartuli. *Urban Infrastructuring: Reconfigurations, Transformations and Sustainability in the Global South*. D. Iossifova, A. Gasparatos, S. Zavos, Y. Gamal and Y. Long. Singapore, Springer Nature: 157–172.

Chatterjee, P. (2004). Populations and political society. *The Politics of the Governed: Reflections on Popular Politics in Most of the World*. New York, Columbia University Press: 41–94.

Chatterjee, P. (2011). *Lineages of Political Society: Studies in Postcolonial Democracy (Cultures of History)*. New York, Columbia University Press.

Chattopadhyay, S. (2005). *Representing Calcutta: Modernity, Nationalism and the Colonial Uncanny*. Oxford, Routledge. **2**.

Chattopadhyay, S. (2012). *Unlearning the City: Infrastructure in a New Optical Field*. Minneapolis, MN and London, University of Minnesota Press.

Cresswell, T. (2004). Defining place. *Place: A Short Introduction*. Malden, MA, Blackwell Publishing. **12**: 127–136.

The Economist (2021). Volunteers are filling the gaps in India's fight against Covid-19. *The Economist*.

Iossifova, D., S. Zavos, A. Gasparatos, Y. Gamal and Y. Long (2022). Introduction: Trajectories of infrastructural entanglement in cities of the Global South. *Urban Infrastructuring: Reconfigurations, Transformations and Sustainability in the Global South*. D. Iossifova, A. Gasparatos, S. Zavos, Y. Gamal and Y. Long. Singapore, Springer Nature: 1–12.

KMC (2017). Official website of Kolkata Municipal Corporation.

Lawhon, M., D. Nilsson, J. Silver, H. Ernstson and S. Lwasa (2018). Thinking through heterogeneous infrastructure configurations. *Urban Studies*. **55**(4): 720–732.

Lefebvre, H. (1991). The production of space. *The People, Place, and Space Reader*. New York and London, Routledge: 323–327.

Lewicka, M. (2010). What makes neighborhood different from home and city? Effects of place scale on place attachment. *Journal of Environmental Psychology*. **30**(1): 35–51.

Ministry of Home Affairs (2011). *Census of India 2011: Instruction for Manual For House-listing and Housing Census*. New Delhi, Ministry of Home Affairs.

News Desk (2022). As Covid-19 cases surge, red volunteers back on Bengal streets to help people. *News 18*.

Routray, S. (2014). The postcolonial city and its displaced poor: Rethinking 'political society' in Delhi. *International Journal of Urban and Regional Research*, Wiley Online Library. **38**: 2292–2308.

Tuan, Y.-F. (1977). *Space and Place: The Perspective of Experience*. Minneapolis, MN and London, University of Minnesota Press.

4 Seasonal adaptations and everyday negotiations

The preparation phase

During the idol-crafting season before the festival, public spaces and infrastructural inadequacies are adapted, adopted, or repurposed by the crafts practitioners of Kumartuli to carry out the practice. This chapter draws on the dynamic between facilities being stretched to breaking point, the community's grievances, and the continuing faith-led consumer demand for clay crafts. Durga Puja in Kolkata usually begins in late September or early October each year. The busiest time of Kumartuli is undoubtedly the preceding days of this autumnal festival. In order to witness the scale of exchanges and the distribution process that takes place here, I selected this time of the year to start the participant observation or 'being there' (Marcus 1995; Hannerz 2003). My fieldwork began days before the annual celebration of Durga Puja in 2017. While I had been to Kumartuli before as part of a small group of postgraduate students on a study trip, I had never witnessed the peak transaction season. As a middle-class, upper-caste woman raised in Kolkata, my gender was perhaps a factor that socially excluded me from being a customer and part of a neighbourhood *Puja* committee party. In Kumartuli, late monsoon rains usually impede idol preparation. The diligent media never fails to inform the eager Bengali community globally about the preparatory stage before the annual festivals. In order to get ready for the fieldwork, I had been attentively following these reports in the local news media.

On a bustling morning in Kolkata, seven days prior to the festival, I visited Kumartuli, equipped with a camera and a notebook. It had been an unusually rainy monsoon, with above-average rainfall, and towards the end of September; the streets of Kumartuli were still wet, and the thousands of footfalls and the clay from the idol preparation made them muddy. The roads leading to Kumartuli from the nearest metro station were busy as anticipated, but the volume of people increased to hundreds, possibly thousands, as I went closer to the inner areas of the neighbourhood. The para-transit system comprising auto-rickshaws (a familiar mode of transit in inner-city areas of Kolkata) was restricted around the Kumartuli neighbourhood that day to reduce vehicular traffic and manage the crowd. Numerous potholes held puddles of muddy water, and the constant movement of motorbikes, vans, and bicycles caused the mud to splatter on the pedestrians, slowing the crowd

DOI: 10.4324/9781003341222-4

and causing confusion. As I walked, I had to stop and yield to the motorcyclists travelling at high speeds to avoid muddy splashes. As a woman who grew up in Kolkata, I was accustomed to muddy potholes, rushing motorcycles, and packed streets during the Durga Puja. However, what stood out about this situation were the people's singularly focused actions on the idol-making cluster. Everyone appeared occupied with their respective activities, but the crowd moved slowly because everyone was in a rush.

The 15 minutes seemed to take much longer, and by the time I made it to the core of Kumartuli's narrower streets, I had fallen behind a sizeable crowd. A group of about 30 to 40 men (and boys, largely teens), laughing and chatting, had come to pick up an idol they had pre-ordered for a festival or *para*-puja sponsored by their locality. They belonged to one of the numerous parties that had travelled to Kumartuli for business. A majority of young people and a small number of middle-aged men from their *para* accompanied all similar *para*-Puja committees when they arrived in trucks or lorries; the older men were haggling last-minute pricing with the idol-makers and porters, as the younger audience cheered enthusiastically (Figure 4.1). Cheerful tunes, often Bollywood music, played through portable stereos are accompanied by their collective chorus of festive notes: '*Bolo Du(r)gga Mai ki . . . JOY*' (Shout out for the Mother Durga . . . *VICTORY*). Laughter, singing, chanting, the sounds of the celebratory drummers (*dhaki*), and anticipation for the year's largest event filled the air.

It was amazing to see the large crowd of people squeezing through the occasionally muddy, unpaved alleys to pick up their pre-ordered idols, even though they were wide enough to accommodate idols that were twice or three times the height of an average person. I kept asking myself, '*Is it hundreds or thousands?*',

Figure 4.1 Distribution of idols: covering the deity to protect from rain; idols on a lorry, ready to leave Kumartuli (September 2017)

and then I understood I was not seeing the same individuals again and over again; instead, there were groups of people arriving and departing with the finished idols while making similar upbeat noises. Despite the mayhem, there was order, which the police's presence helped to maintain. In order to manage the sizeable crowd efficiently, the Kolkata Traffic Police were in charge of both the vehicle and pedestrian traffic. Police patrolled the major thoroughfares and the inner lanes to maintain traffic. The roughly eight-feet-wide streets of Kumartuli are too narrow for trucks. These are parked on the broader Rabindra Sarani, also known as Chitpur Road, and are open only to privately owned and idol-carrying vehicles at this time (Figure 4.1). Idols must be pulled from the workshops to the trucks on hand-drawn carts by groups of porters.

I waited, I watched, and I walked through the streets, witnessing conversations and distribution activities in Kumartuli. It was just past lunchtime. There were groups of porters finishing their hastily prepared lunch of a plate of rice and some curry by the river and washing up. I later understood that most of them had come to Kumartuli as seasonal workers just for these few days; otherwise, they would not be staying in the open by the river. There were various groups of people, some of whom were cooking on a makeshift kerosene stove, while others hurriedly approached the river to take a quick dip before eating, and, still, others rested briefly at the street corners before setting out again with renewed strength to pull the heavy idols. Some groups were rushing back to their stores out of concern for losing a potential clientele. The porters carried hand-made carts and ropes to pull the idols to the trucks waiting on the main road. One 20-foot-tall Durga idol was being pulled by about six porters, and there were still hundreds more to load into trucks. I studied them carefully as they pulled the intricate sculptures; the man at the very end recited a chant for the others to follow as they pulled. I wondered how much effort it must require to haul an idol thrice the human scale.

I entered one of the narrower alleys. All the shop front shutters had been opened up to maximise entrance. There were workshops on both sides, some were almost empty, and a few artists were still putting the final touches on their idols (Figure 4.2). Young photography enthusiasts were flocking outside these workshops to take some magnificent pictures. Quite a few young women dressed in traditional, red-bordered *saris* were posing with the unfinished idols for the cameras. Some would be published on the media, perhaps social media, portraying the ordinary and the mundane 'everyday' of Kumartuli. Some images would make it to magazine covers illustrating women's empowerment symbolising the Mother Goddess of *Shakti* (power) through the Bengali Hindu deity of Durga. The female deity of Durga represents empowerment, and the immense entanglement of the festival with the average Bengali's lives firmly embeds the idol-making practice in the local culture.

Durga Puja brings together a range of emotions and anecdotes associated with the worship of idols. Folk stories suggest that Durga Puja is the celebration of homecoming. Durga, associated initially with the image of a daughter in folk culture, visits her parental home in Bengal, accompanied by her four children from her marital house in *Kailash* (imagined somewhere in the higher snow-capped reaches

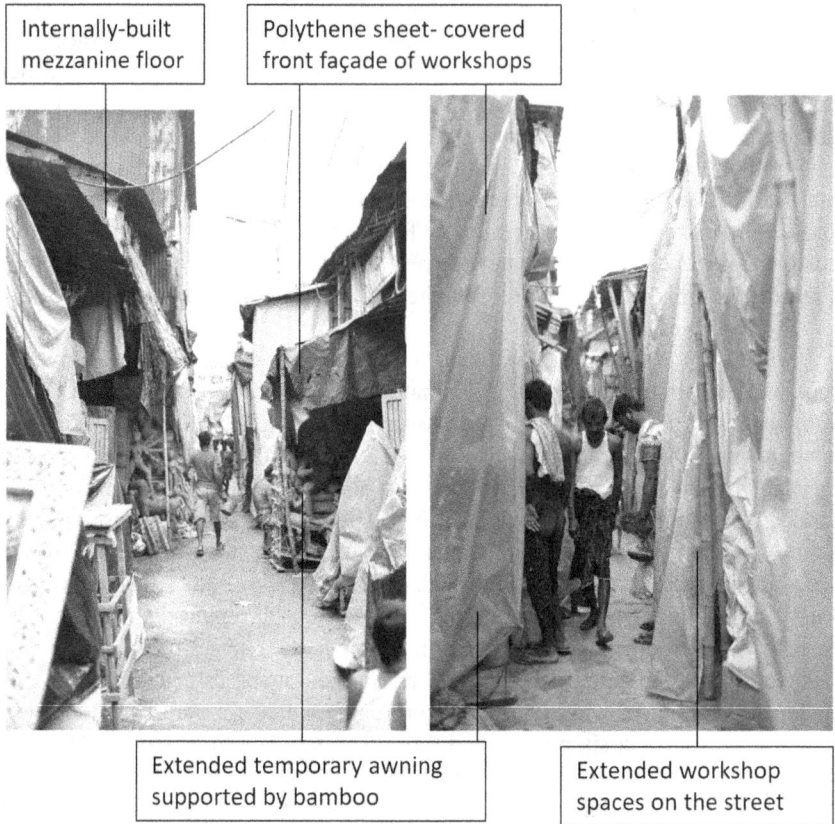

Figure 4.2 Condition of the road in Kumartuli: during the late monsoon, September 2017

of the Himalayas) (Bhattacharya 2007; Bean 2011). Each of the four mythological children is God and Goddess themselves and, along with Goddess Durga, is gifted with a significant aspect of human livelihood like wealth, prosperity, education, arts, might, and strength to kill the demons (interview: Bagbazar 2018). Some rituals performed during the Durga Puja are related to regional folk customs performed while celebrating the homecoming of a married daughter and her family. The festival traditionally brought occasional treats like confectionery, flowers, and new clothes, which an average person struggled to bring home daily (interview: Kumartuli 2018). Hence, the public sentiments towards Durga Puja are more than ritualistic offerings of prayers; it brings about general happiness, family gatherings and festivities, and, more importantly, includes participation from all social spheres.

I remembered some pictures in local newspapers and social media and imagined youngsters of similar age group taking them. Some *kumars* seemed perturbed by the buzz caused by these photographers, and I was warned earlier by local friends that they often use harsh words to disperse such a crowd. I saw a group of such photography enthusiasts walking away in disgust from one of the busier work-shops. I followed them to the end of the alley and took a turn in an attempt to see the workshops more closely. This part of the street was less crowded; I did not take pictures, so I was not shooed away. I asked permission to watch and witnessed the artist put the nose ring, add the intricately embellished crown to the idol, and secure it with adhesive. It was a smaller idol, perhaps for a family puja. Hence, his customers would probably visit a couple of days later to pick up the idol when Kumartuli is a little less crowded. There appeared to be no rush for the domestic celebrations; the festival would begin in seven days. Only the neighbourhoods with higher budgets for revelry open their pavilions a few days before the actual festival for the public. Notable politicians and celebrities inaugurate these intricately built *pandals*[1] (pavilions sheltering and showcasing idols) to bring in patronage. Family-organised pujas, however, follow the traditional routine and rituals.

Within five days from the end of Durga Puja, another Bengali puja for the Goddess *Lakshmi* is celebrated. This celebration is not huge. It is mostly a house-hold festival, but idols are required. Besides, all Durga Puja committees perform *Lakshmi* Puja to mark the end of the season. Hence, smaller workshops were stacked with *Lakshmi* idols, and artists were busy finishing them. Perhaps they were preparing to finish the work before Durga Puja commenced. Shops were sell-ing decorative and traditional paraphernalia required for the festivals at both ends of the neighbourhood. These shops cater to the numerous customers buying idols from Kumartuli and provide additional customary ritualistic embellishments for decorating a festival pavilion; some decorations are made from *shola* (piths, mostly roots) and some from *zari* (interview: Kumartuli 2017, 2018). As evening fell, the lights of the shops selling decorative items for the festivals dazzled with the glittery merchandise hanging in front of these smaller shops. There were multicoloured cut-outs made from polystyrene sheets, garlands of synthetic flowers, and other colourful festive insignia. Women appeared in the neighbourhood as evening fell, not in large groups but alone or in pairs. It is most likely that while the men visited Kumartuli for their *para*-puja, the women came for predominantly domestic festiv-ities. I saw ladies haggling at a bargain and buying these items for their households. These shops cater not only to the more prominent festivals but also to the ordinary Bengali families, who come to purchase ritualistic embellishments for a festive makeover for their homes. Practices allied to idol-making coexist and complement each other in the Kumartuli neighbourhood during the festive seasons.

A few other businesses situated on the outer lanes of the neighbourhood are busy during this time of the year. On the wider roads, shops and temporary sheds are set up where items like ropes, bamboo, heavy-duty polythene, and canvas sheets are sold. These items are required for holding the idols in place during the transit, shad-ing, and similar purposes and for constructing temporary tents or *pandals*. These businesses, although independent, are hugely reliant on the idol-making industry. Some of them have permanent shops in Kumartuli near the riverfront but have to encroach upon the streets and set up makeshift tents to make room for their excess stocks during the festive seasons. Among the several bamboo-framed and canvas tents set up during the festive season, many are used to house the seasonal migrant population temporarily. These tents temporarily accommodate large groups of *dhaaki* (festive drummers) who travel to Kumartuli to be hired by the Puja committees. Like the porters, *dhaakis* are a seasonal workforce dependent primarily on autumnal festivities. The *dhaakis* arrive at Kumartuli from their village residences in large groups before Durga Puja and leave after *Lakshmi* Puja or even *Kali* Puja. They are usually hired in groups (of kin or families) of mostly men and boys in Kumartuli to accompany the idols. They play the *dhaak* (drums), an essential ritual for most Bengali Hindu festivals. Once hired, they would stay at the Puja venue for the festival and return to Kumartuli at the end of Durga Puja to regroup and leave for their village homes.

Hundreds of men from rural Bengal come to Kolkata and the neighbouring urban areas to pursue seasonal jobs. I witnessed large groups of such men and young boys waiting to be hired on the outer streets of Kumartuli in the afternoon. The seasonal workforce that travels to Kumartuli in preparation for Durga Puja and related autumnal festivities belongs to mostly marginalised castes from Bengal but serves the most significant Hindu festival (Chakrabarti 2022). Scholarly understanding of Durga Puja as a religious practice and Kumartuli as a craft neighbourhood must be located against the complex backdrop of the growing commodification of a cultural craft, as well as an understanding of how the seasonal workforce accompanied by festive drummers and associated caste-based professions and other networks have evolved over time to facilitate these operations of practices.

The Kolkata Metropolitan Development Authority (KMDA 2009) suggests that there were in total 298 business establishments in Kumartuli during their survey, of which 166 belonged to the idol-makers, and the rest were commercial units, including 51 pith (Bengali: s*hola*) merchants, and at least 39 of the rest sell outfits, items of rituals, and jewellery for the idols. The others are mostly roadside tea sellers, small groceries, or corner shops selling everyday items. Some of these shops were established in the early 20th century and are still run by the same family [Interview: Kumartuli 2017, 2018]. The neighbourhood, less than a square kilometre in area, accommodates people and practices much more than usual during the preparation phases of Durga Puja each year. In order to provide for such a massive influx of people, numerous makeshift arrangements change the regular spatial character of the place. Workshops extend onto the alleys, shops set up temporary shed extensions in the front, and the streets are social–cultural spaces of interaction and business transactions (Figure 4.2). The lanes are even temporarily only used for performing everyday practices such as cooking, cleaning, and resting or sleeping for people who form part of the seasonal workforce involved in the wider practices closely associated with the idol-making industry.

Adaptations, accommodations, and negotiations

Idol-crafting in Kumartuli is interwoven within the everyday lives and practices of the residents. This section builds on visual documentation and the narratives of the places of production within the homes, workshops, and streets of Kumartuli by the participants of the photo-study. Away from the seasonal adaptations and appropriations, Kumartuli is home to a permanent group of residents. Seasonal migrants, while constituting a significant influx during the autumnal festival preparation, are not everyday residents of Kumartuli. Hence, this photo-study only involved the everyday residents of Kumartuli because, first, this was conducted during the off-peak season to increase participation, and, second, to gather views of everyday negotiations rather than the seasonal ones. These pictures reflected the frugality in the everyday lives of the residents of Kumartuli in general. Family history, stories of struggles, aspirations, and attached emotions to residences were a theme in these sets of pictures. In the first part of this section, different perspectives of the residents' views of their neighbourhood have been reflected through their ideas of the social, cultural, and the built-environment of the places they dwell and work in; these perspectives have helped to construct the meanings of these places.

Everyday adaptations, modifications, and negotiations are both personal and a result of collective actions. Through the photographs, participants were interested in showing me (the researcher) and perhaps their distant consumers glimpses of their everyday reality. For example Figure 4.3 (R7–6, R7–14)[2] shows the condition of a workshop. Low-light conditions and the interesting use of the tree trunk as part of the workshop (Figure 4.3, R7–6) are regular pictures. The spaces in Kumartuli are at a premium, and every corner is essential for the production and storage of items. The images had been clicked during the day; yet some were quite dark and had to be discarded. The stories, however, are common in these pictures, and all related to the everyday struggles with cramped spaces, low-lit conditions, and dampness.

Figure 4.3 Everyday realities within a workshop/residence [R7–6 (tree-trunk used as an integral part of the workshop), R7–14 (working space on the workshop floor), R7–16 (kitchen), and R1–8 (old roof structure built from bamboo, fitted light)]

Glimpses of indoor spaces include the view of a workshop with the owner in action (Figure 4.3, R7–14). The only picture (Figure 4.3, R7–16) of a kitchen with a similar damp and cramped condition is portrayed. The participant strongly expressed an attachment to the kitchen:

> My kitchen is small and dark, but it is my own place; I spend most of the time here cooking and cleaning and cooking again for my husband and children.
>
> (Interview R7)

Although it is the only picture of a kitchen (Figure 4.3, R7–16), this portrays the gendered nature of the everyday practices in Kumartuli in more ways than one. None of the men in the study have taken pictures of their homes' or kitchens' interior spaces. Therefore, they could not be questioned about everyday household practices.

Figure 4.3, R1–8 shows the picture of a roof with a precariously fitted light and the roof structure built from bamboo around 40 years ago. While one of the participants highlighted this condition, a similar low-lit and precarious roof supported by a single bamboo pole can be seen throughout the neighbourhood. Also common are the lack of storage and temporary display of idols and statues for sale. The reasons for carrying on with the old and deteriorating structures are not only financial but sentimental and the changing seasonality and growing demand of the idols as well. Some workshops struggle to keep up with the growing demands of other sculptures than religious idols beyond the traditional seasons, while others have sentimental values attached to the workshop buildings once built by their forefathers (Figure 4.3, R1–8) (interview R1, R2).

Old buildings (Figure 4.4, R4–2) in Kumartuli are symbols of sentiments and family histories. However, due to ownership issues, these buildings are in a state of disrepair and a potential safety hazard due to exposed electrical wires. Similarly, the building in the picture (Figure 4.4, R3–7) is an old workshop. Due to the lack of family support, and the discontinuation of the caste-based profession, this workshop, once belonging to a well-known artist, lay empty and ultimately derelict (interview R3).

The cramped spaces, frugal building construction, and materiality illustrate the everyday challenges faced by the families involved with idol-making in Kumartuli. However, the residents believe that

> This is my 'motherland'. I was born here. Other people might not like it, they might think this is congested, but I have grown up here. This place . . . I like very much. Now, you all might think it is broken, but is it really?
>
> (Interview, Kumartuli 2018).

The struggle with space within the workshop presses the residents to extend their practices beyond their thresholds. This is a common sight on the streets of Kumartuli. However, many participants have pointed out that these actions are impacted not only by space constraints but also due to lack of planning, off-season

Figure 4.4 Grievance about the workshop and buildings and workshops that overtake the streets [R4–2 (old buildings, potential safety hazard), R3–7 (old workshop, now empty and hazardous), R8–19 (straw disposed of the street), R10–4 (street used as a part of the workshop)]

works, and overall tolerance for these actions across the neighbourhood (interview R1, R3, R8, R10). Lack of planning and off-season work is somewhat related:

> If you buy extra straw during the monsoon, it is bound to rot and would end up on the streets (Figure 4.4, R8–19), but in most cases, you have to buy it because you have to complete orders and ship them overseas sometimes— that takes time to reach the destinations.
>
> (Interview R8)

Additionally,

> Kumartuli was not as crowded—had less traffic and fewer people before, so it was normal to extend onto the streets for space (Figure 4.4, R10–4); now everyone 'pushes' each other with their elbows.
>
> (Interview R10).

The interview with R10 was particularly interesting; I walked around with the participant to help use the camera. On this walking tour, the participant narrated the history of the places of Kumartuli that he intimately felt familiar with. The tales of different landmarks of Kumartuli were unfolded to me through the eyes of a participant who also narrated the stories behind certain everyday practices and rituals. I have combined some stories from R10 and other photo-study participants to retell them coherently.

The other reason cited by one of the participants for using the streets was because of sunlight and air. The unfinished idols (or smaller parts) are often placed on the roads for drying (Figure 4.5, R3–9, R9–5). Also, clay-mixing work cannot be done within the workshop for similar reasons (interview R3). This practice results in narrowed or blocked streets for part of a day (Figure 4.5, R10–13), but the clay is mostly removed from the streets after mixing in adequate quantity for the day's work (interview R10).

> The structure of the workshops with foldable main gates opening onto the streets (Figure 4.5, R10–11) helps natural light to come in. Most of the workshops were built before the area was electrified, so no (electric) fan or light was there. Back then, we used to work in the same way as we do now because we learn our skills from the previous generations.
>
> (Interview R10)

This suggests that the practices of idol-making are as much dependent on the spatial arrangements of the workshops as the underpinning technical know-how or

Figure 4.5 Everyday practice on the streets of Kumartuli [R3–9 (unfinished idols on Banamali Sarkar Street), R9–5 (parts of idols left on the street to dry at Gokul Mitra Lane), R10–13 (clay mixing on-road), and R10–11 (foldable shutters opening on the street)]

competencies of the practitioner. The insights of everyday practices received from R10 are manifold and construct the complex narrative of the places in Kumartuli. Due to his age and experience, he has witnessed the changing production patterns in the recent past and believes in continuing to practise age-old norms and rituals. However, he also embraces a changing demand that urges him to adopt newer seasonal pressures of idols, with younger generations within his family participating in his business. He carries on the legacies of the past through storytelling and wishes to train the younger artists to follow his traditional practices.

Photographs presented in this section portray a range of different scenes, some pictures of buildings or the built environment and others of people that inhabit or work in these spaces. Some photographs are reflections of a combination of the themes; however, the overarching components of all photos are reflections of the sentiments and resentments of the residents of the neighbourhood and important tools to analyse people's voices—the underpinning power structure and the political economy of the idol-making industry as a whole. Participatory research is a collaborative and engaging method of understanding not only the everyday lives of the participants but also their relationships with their immediate surroundings—people, things, and location. This allows for making informed decisions and actions instead of perceiving things from the researcher's or the decision-makers' perspectives (Pink 2012). Moreover, multiple meanings and emotional significance of a place for the users, passers-by, or residents and the reasons for including certain places instead of others by the respondents make the technique quite useful for psychological insights into the people's perception of the place. This helps in a socio-spatial construction of a place, negating the duality of understanding the informal urban setting of the global South representing both ordinary and extraordinary spaces through visual imagery and narratives.

Place attachment, meaning emotional bonds to a place in environmental psychology literature, is often related to the social dimensions rather than the physical dimensions of the place (Lewicka 2011). The sense of place has been viewed predominantly as a '*social construct*', a '*product of shared behavioural and cultural processes*', rather than '*rooted in physical characteristics of the setting*' (Lewicka 2011, pp. 214). In the case of Kumartuli, while residents are attached to the place due to their social ties and relations, the photographs suggest that there is a construction of meanings with the built environment and the social infrastructure. The photographs reflect as much on people as on the meanings of the spaces of cultural production. The results of the participatory photo-study reaffirm the strong place-attachment of the residents and call for further exploration of the processes involved in making the meaning of such places for a better understanding of informality.

The reason these images are important in constructing the meanings of place in a *basti* neighbourhood of the global South is not only the apparent visual perspective of the neighbourhood—images of poverty, frugal means of living, and dilapidating building. Meaning, however, lies in the interpretation of images from the perspectives of the residents, who have constructed profound and intimate emotions in these places of production of a cultural and economic product of immense significance to the people and government of the city and state. Although the economic functions identified by the state government within the neighbourhood call for the drastic and unsympathetic portrayal of global standards, the residents' perspective of 'showcasing' the smaller details of their everyday lives to the wider public was mostly unheard and overlooked in the KMDA-proposed plans of redevelopment between 2008 and 2011.

Infrastructural disrepair and hopelessness

Place is perceived as a viable object from the user's perspective, irrespective of formal or informal nature of land and building ownership within a neighbourhood. Applying a place-based approach to understanding practices within the complex flows and networks of the idol-making industry uncovers unheard voices. The participants' emotions and reflections are presented through the photographs they have taken. To narrate the stories of different residents in this study, I have grouped the pictures from a macro scale to the micro, from the outside to the inner spaces of the neighbourhood, and from the wider problems to the more specific concerns. Hence, the stories regarding the outer areas of the neighbourhood and the networks like roads, the river, and the rail are sequentially followed by the scenes of the inner streets and the everyday lives of the residents as part of the community. This section concludes with photographs reflecting grievances and discontent about the changing and deteriorating conditions as well as the everyday struggles of living in Kumartuli, as described in the participants' photographs and supporting interviews.

In this regard, the 'places' in a neighbourhood are often constructed by the residents—the informal spaces of the neighbourhood form places of production of economic functions. Different scholars have defined places from different disciplinary perspectives, and it is important to note why and how a neighbourhood forms the basis of place-based perceptions of the residents. Friedmann (Friedmann 2010) notes that neighbourhoods, even 'slums' with inadequate infrastructure and low housing, are the places of people's everyday lives that they are familiar with and connect to. He writes,

> [N]eighbourhood is cherished for very different reasons: because it has places of encounter where people reaffirm each other as who they are or comment on the day's events; because life has a certain rhythm with which all are familiar and to which all expectantly look forward; because there are places that are 'sacred' to the people; and because there are special places of gathering where events important to the community transpire. It is this rhythm, these repetitive cadences that are always the same and yet a bit different as well, like a seasonal festival, that is a measure of a neighbourhood's vitality.
> (Friedmann 2010, pp. 162)

Through the photographs, respondents have reflected on the disruptive environmental conditions and everyday hazards that illustrate poverty and frugal means of life. However, none of the interviewees have directly stated that they want the government to 'upgrade' or 'redevelop' this area because the conditions are hazardous enough. Instead, they said regular road works, fixing drains, and cleaning the garbage would benefit them a lot. Further, they considered altering and mitigating daily habits that cause such hazards and called for strengthening local governance.

In the deliberative workshop I conducted in February 2022, attended by the residents of Kumartuli and policymakers, community representatives repeatedly reiterated the importance of repair and maintenance. Empirical literature from across the global South in Urban Studies echoes the need and implications of repair and maintenance of existing infrastructural systems and incremental changes to provide access to marginalised communities (Simone 2004; Bhan 2019).

In the interviews, for example, respondents highlighted the proximity to important amenities such as schools and hospitals and good connectivity as reasons for residing in Kumartuli. However, in the photo-study section, I have illustrated and highlighted that the residents of Kumartuli have more reasons to not only stay in but also thrive in Kumartuli, despite the failing infrastructure, the auto-constructed[3] buildings, and constrained spaces. I observed a duality in photographs such as attitude towards neighbours—some respondents valued their neighbour's support and friendship, while others reflected on habits that hindered their friendship and coordination—such as garbage disposal in inappropriate areas and careless placement of raw materials in front of workshops and on the streets and further contestation with the delivery of services for potential clients.

Analysing the photographs and the subsequent interviews were important tools for understanding the power structure and agency among the residents. I used participatory photography in this study to put voices to the marginalised residents of a slum neighbourhood. I tried to engage participants from all groups, such as younger, middle aged, and older artists and women and young adults in the neighbourhood, but I could not find famous artists to participate in the study. The photo-study involved a cross section of the residents and mindfully presented as many views as possible. Still, there could be an analytical issue with the photographs the participants took: do the photographs only reflect the voices of the subaltern and speak only about power relations? Do these only reflect the discontent and aspirations, or do they illustrate the residents' needs and attachment to the place? There is a slight danger to studying photographs taken by only the struggling residents; however, the place-based approach is suited because there was a mix of residents who reflect their place-attachment and memories that are shaped by these places.

In the narratives that follow is the description of the photographs and accounts of the same given by the participants of the photo-study. These reflections helped me to provide multiple views of certain places and reimagine the neighbourhood.

> I grew up here, bathed and learnt to swim in the river. The river is an integral part of the daily life of Kumartuli: people bathe and perform rituals here (Figure 4.6, R5–1), and the bamboo and clay come through the river. Kumartuli became 'successful' because of the river—you know, the Ganga mati (clay from the bed of the river Ganges).

Nevertheless, he reflects on the river more with his childhood memory:

> [E]very morning my father, after moving here from Bangladesh, used to dip in the river and worship in the nearby Dhakeshwari temple. Earlier we used to bathe in the river, you know, with brothers and cousins while growing

Figure 4.6 Top: riverfront and the *ghats* [R5–1 (riverfront), R8–12 (Strand Bank Road)];
bottom: railway track parallel to Strand Bank Road and the tea stall by the level
crossing [R5–13] and road (tram line) [R4–5 (an open squatting type 'public'
urinal and small roadside temple)]

up. . . . I do not like how people throw garbage near the river now, but it is
part of life. . . . I think.

(Interview R5, a middle-aged artist, who lives by the river)

Although the river is one of the common themes in photographs, every participant
had a different way of constructing the meaning of the river and the riverfront.
Some views were of pride in the heritage of Kumartuli, while others said,

[Y]ou will only find bamboo stacks in the Kumartuli Ghat, and not on many
other ghats (riverfront embankments) of Kolkata if you see the bamboo or
the clay-boat, (. . . and) you know you are in Kumartuli.

(Interview R10, elderly artist)

In addition to being a transport linkage, the riverfront is also a place of respite for
the youth. One participant took pictures of the riverfront where he and his friends
go for an evening hangout; groups of friends and couples go for leisurely walks on
the Strand Bank Road while enjoying the pleasant breeze in the summer.

[B]ut we are often distracted by the odour and dirty (unhygienic) conditions due
to garbage dumping, he added with environmental concerns (interview R8, young

artist, works with his father). However, a lot of the floating debris on the river are refused flowers and remnants of everyday religious practices that take place on the *ghats* and have been consistently disposed of in the river for 'ages' (interview R5). Residents' place-making activities within informality contend many of dominant ideas circulating these places (Lombard 2014). The photographs express much differentiated views that often oppose negative portrayals of certain activities. Hence, while the riverfront is a place of respite for the residents, a local discourse about health and hygiene compromise due to religious rituals and floating debris from idols distorts the image of the place. However, that does not change the convenience of being close to the river for idol-crafting or the fond memories of the riverfront.

Likewise, pictures of the railway track and the level crossing were taken to illustrate not only transport linkages but also as places to meet friends at a corner tea stall by the riverfront.

> We used to play by the railway track or have a chat afterwards during the summer afternoons when travins came every hour. . . . As soon as we heard the horn, we would move out of the track and wave at the passengers on the train. Later on, I realised how close the Shobhabazar station is, and how convenient that is for travel.
> (Interview R6, middle-aged artist of average acclaim)

For another participant, the tea stall and the shed (Figure 4.6, R5–13) by the level-crossing were places where he spent evenings during his youth (interview R5). The tea stall set up more than 30 years ago as a local initiative for serving the residents remains and continues to be an essential part of the social life of many residents of Kumartuli. The tea stall is by the Strand Bank Road, outside the redevelopment project proposed earlier. The frugally self-built tea stall, encroaching the road and pavement, made of bricks and bamboo and a temporary tile shed, might lose business or might have to move due to the 'world-class' tourist hub aspiration of the local government and the rising real estate value of the adjoining areas of Kumartuli—a result of the marginalisation, gentrification, and spatial shift due to urban redevelopment projects in India (Ghertner 2014).

Another recurring theme of the pictures was the road linkages in and around Kumartuli. The photographs included wider roads on the periphery of the neighbourhood as well as narrower inner alleys. The reasons cited for these pictures were mainly the conditions of the roads and how that affects the everyday lives and practices of the residents of the neighbourhood. For example one resident said,

> I think the condition of roads in Kumartuli could have been far better, take Rabindra Sarani, which is in such good condition, then why can't the neighbourhood streets be better (?)
> (Interview R3, male artist, relocated to Kumartuli 2)

However, the same Rabindra Sarani was pictured with the 'open public squatting-type urinal' (Figure 4.6, R4–5), citing the foul smell and visual hazards of the scene. Also, he said,

> [T]here is a public toilet (pay to use), just around the corner, but because people have to pay to use that, they use this instead. . . . Don't you think this is smelly and dirty? Moreover, it is just by the entrance to the neighbourhood, so what picture does it give about us?
>
> (Interview R4, young adult)

These streets are deeply rooted in the residents' everyday life and show how the people interact with their immediate environment, with constant making and unmaking of spaces. The street in Kumartuli is a spatial, social, and cultural entity, within both personal and public domains, characterised by activities with substantial overlapping in different temporal frames. For example a grocery shop, whose threshold or window is on the street, has customers gathering, waiting, and mingling in front. The open workshop shutters also portray the street, which serves as a space for interaction between a passer-by and the *kumar* in his workshop. The narrow street is not only a means of transporting and distributing the products from Kumartuli but also a space of interaction between neighbours, a passing client who might be impressed by the artistic skills of the *kumar*, or simply an extended space for production.

The streets of Kumartuli are spaces of social and cultural production. Also, unique visuals, sounds, and smells are generated through the everyday practices of the residents. These multi-sensory experiences are quintessential realities constructing meaningful places within cities: what is seemingly unpleasant to a passer-by is of multiple meanings to everyday users. The sound, smell, and visual imagery support the construction of these places and lend extraordinary meanings to the seemingly ordinary places of practices.

However, the multi-sensory experiences of every day and the attached sentiments of the streets of Kumartuli are overridden by the participants' grievances due to the degrading condition of the infrastructure and public services. The road network is particularly essential in this regard because it plays a significant role in procurement and distribution. The inner streets were seldom repaired. I was told by a resident (R8) that there was a waterspout set up, and the road was dug due to that more than a year before I visited for fieldwork in 2018. As of the dates the photos were taken, the street was not repaired. Due to this, he (R8) and his father face daily difficulty and sometimes irreparable damage to finished clay idols while transporting them and hold the local councillor responsible for not listening to their grievances. He (R8) also partly blames neighbours for the lack of concern for health and hygiene in the neighbourhood while admitting there is a lack of facilities to dispose of the regular waste from the workshop. On the contrary, others reflect on the complaints of residents regarding infrastructure and how these degrading infrastructures lead to hazards. Typical street scenes show the overarching sheds of workshops. The grievances related to the infrastructure came up as a theme from questions related to struggles of every day and letting their distant consumers know of these. Again, an outsider may stigmatise the locals by these visuals of the slum, whereas these are results of negligence and lack of public works by the local authorities.

I observed general discontentment with the political leaders and the government (interview, Kumartuli 2017, R3, R8). Since the political party in power had promised 'total reform' (*paribartan*, in Bengali), people had voted for them, hoping for good governance. However, most people have expressed their discontent, which has increased because of failed promises (interview R3, R5, R8, R10). According to the residents, certain areas of governance would have to be proper provisions for garbage disposal and better conditions of roads and drainage networks. Residents pointed out that road repair works had not been done since 2010. They now believe that all politicians are the same but might take a chance to vote for someone new in the next elections (interview R3, R8). The grievances regarding public works, services, and infrastructure were mainly targeted to the local governance mechanisms and elected representatives, which were restated during the deliberative workshop in 2022. In addition to the discontent due to the lack of services and overall degrading environmental conditions, the participants have shown general dissatisfaction due to the lack of initiatives by the current government in the area. These include services like garbage collection, overhead electric supply wires and posts, and maintenance of roads.

Social cohesion, coordination, and competition

The idol-making cluster in Kumartuli is supported by an array of other ancillary businesses, including jewellery and outfits production for the idols, household scale hosiery, and cardboard-box-producing industries, which strengthen the businesses of the idol-making cluster and employ an additional workforce from the residents. These smaller but important economic activities have also been reflected through the photographs. A participant (R7), who is female and family member of a *kumar*, is also aware of the immense support that the ancillary trades provide to the idol-making practices. She said,

> [B]ecause I have grown up here in Kumartuli, I understand that every other profession than idol-making forms an important part of the industry. For example, among the ancillary industries based in (the) Kumartuli neighbourhood are garment making. One of the seasonal supplies for idol-making, the hair made of jute fibres (Figure 4.8, R7–9), is ferried by the occasional seller who only comes to Kumartuli neighbourhood and no other neighbourhoods of Kolkata. When I was younger, I used to fold cardboard and make boxes (Figure 4.7, R7–8) and earn some pocket money, very little, but I was happy!
> (Interview R7)

Building on interviews and analysis of photographs, I argue that most of such smaller enterprises are part of a wider informal economy struggling with the lack of capital and do not have large profit margins. Displacement of the practices or the redevelopment project would threaten the existence of smaller-scale enterprises that essentially depend on Kumartuli's idol-making business. Similar thoughts

Figure 4.7 Top: importance of support network, ancillary industries [R7–9 (supply of hair for idols) and R7–8 (small-scale cardboard box enterprise)]; bottom: important landmarks and notable workshop [R10–12 (Madan Mohon Temple Rasmancha, Kolkata, where a seasonal fair is organised) and R7–12 (workshop of the famous artist Dhananjoy Rudra Pal)

about the redevelopment project or relevant displacement fears were reflected in other Interviews.

> [B]ecause of the nature of our job, it was not possible to continue from some-
> where new. We work with shola (which is usually white in colour), that will
> get dusty fast.
>
> (Interview, Kumartuli 2018)

This statement refers to a concern from members of the *shola*-crafting community, which is largely an associated practice of idol-crafting and various Hindu religious ceremonies. As discussed in the previous and following chapters, the *shola* merchants based in Kumartuli were faced with similar concerns of displacement due to the redevelopment project. Most of them are relieved to be able to continue their businesses from their old premises due to the failure of the redevelopment plan.

> The clients, they do not know about any new place, they know Kumartuli. Ear-
> lier, we had mostly clients from Kumartuli only. Kumartuli is well-connected
> to everywhere. However, we also have customers from distant places, from
> everywhere in India, such as Bihar, Odisha, Assam. This establishment (his

Figure 4.8 Top: the 'micro-culture' on the streets of Kumartuli [R10–14 (raw materials left on the street) and R5–8 (encroachments on the street)]; bottom: the importance of cooperation and coordination of neighbours [R3–2 (cooperative office of artisans) and R6–6 (respondent's extended family home)]

shop) is here for more than a hundred years, from my grandfather's generation. . . . We also get customers who come to visit Kumartuli during the peak season, and we have a lot of sales then. Moving is not a question.

(Interview, Kumartuli 2018)

Shola merchants, too, have a clientele spread over a wider geographical area than Kolkata because of religious and cultural rituals in eastern and northeast Indian states being similar to that of West Bengal as well as the presence of large Bengali communities in these states. These statements, while supporting the argument of interwovenness of the businesses in Kumartuli, also highlight the seasonality and place-identity.

Kumars also spoke with similar fears of displacement, but through their statements, suggestions of a growing clientele despite the ongoing social–spatial changes are evident. For example the following excerpt highlights the perception of the artists of the centrality and the branding of Kumartuli as the thriving idol-crafting hub.

[C]lients do not go to other places; they come to Kumartuli only. They only know about this place in Kolkata. All old clients come here, and new customers come here because of the name and fame of Kumartuli.

(Interview, Kumartuli 2017)

More importantly, statements such as the one above highlight residents' urge to remain and continue to operate their businesses in Kumartuli. Also, the smaller businesses employ several Kumartuli residents during the off-peak seasons of idol-making; hence, resettlement might mean social, cultural, and economic challenges too many, maybe even shifting them to the margins. Due to displacement, some residents might lose their livelihood, while some businesses would lose their skilled workforce. Residents say they cannot learn a new skill to earn a living in their forties and fifties (interview R1, R3).

Jewellery-making is a crucial ancillary craft that is directly related to the idol-making practices. Among other places in Kumartuli, his (R9) uncle's jewellery-making shop is a place of fond memories, where he learnt the skills and took up his uncle's profession.

> This shop is the place where I spent the summer afternoons watching my uncle make the jewellery, then I learnt to make it. . . . And in winter mornings and afternoons, I played cricket with my friends on the street in front. . . . We still play on the street sometimes.
>
> (Interview R9, young *shola*-carving artist).

According to the participants, other factors that make the neighbourhood unique are important buildings and landmarks that have been built over time and add to the 'branding' of Kumartuli. The sentiments related to the place have been reflected through the pictures of various temples, heritage buildings, and structures. These include locally famous residences and temples (Figure 4.7, R10–12) built by local *zamindars* and *babu*s and workshops of famous artists (Figure 4.7, R7–12). The residence of one of the members of Shobhabazar Rajbari still adds to the list of heritage structures around Kolkata prepared by the Kolkata Municipal Corporation. However, this structure is in such disrepair that its history might not have been obvious to me before the participant pointed it out. He (R10) also recollected memories of his youth with the building and the temple (Figure 4.7, R10–12) when they were in better condition and still in use and were places of seasonal fairs and festivities. Such recollections and narratives of social and culturally significant places that respondents collectively identify with suggest enduring accounts of familial attachments.

Participants described these buildings and their significance towards the neighbourhood itself. Through these photographs, participants have discussed the neighbourhood's history, the development of the allied crafts, and their attachment to these. Celebrated family-run Rudra Pal's workshop was the subject of a few photographs of the participants. One participant also noted that he was famous throughout Kolkata (Figure 4.7, R7–12). The importance of the workshop is related not only to the location at a significant street corner but also to the cultural legacy of the artistic style of *Rudra Pal* that continues to be associated with his son and nephews. He was the artist who crafted prize-winning idols of older *baroyari* Puja committees like Singhi Park, Ekdalia, 21 Pally, Ballygunge Sarbojonin, and many more. [R3] says,

> Most people in Kolkata would recognise the idols of Mohan Banshi as one of the famous artists of contemporary idol-crafting. I use it (the workshop) as a landmark when I provide the direction to my workshop: like the third house on the left on the lane opposite to Mohan Banshi Rudra Pal's workshop.
>
> (Interview R3)

The description he attached with this picture and the pride of place this workshop studio holds in Kumartuli suggest that, somehow, the residents associate closely with particular places that define the neighbourhood and illustrate creativity and uniqueness.

Regularly, tourists come to visit Kumartuli and have a look at this workshop, if not anything else (Interview R7). Also, the temples reflect not only the place attachment but also the emotional and religious beliefs and resilience of the residents as part of their everyday lives. R7 said that she goes to the *Dhakeshwari* temple every day because God gives her the strength to get through difficult times, and her mother used to pray in the same temple as well.

> [T]he deity is very sacred; she was brought from Dhaka and continues to be worshipped.
>
> (Interview R7)

Neighbours play an important role in the social support system of the neighbourhood and social capital. The photographs and interviews have reflected profound sentiments about neighbours who have resided next to each other with their families in Kumartuli for generations. Some of them have been living together and working there as a family unit for three or four generations and continue to do so on the same premises. All of them have a close-knit integrated socio-spatial structure and are part of each other's everyday relations. As noted by Friedmann (Friedmann 2010), neighbourhoods are essentially places of 'social–spatial' production, where the neighbours interact with each other regularly and thrive. A common thread running through the photographs were visuals of their neighbours, their workshops and residences, and everyday hazards caused by the neighbours. This supports coordination and cooperation among peers in an otherwise marginalised neighbourhood. Residents in Kumartuli are generally close-knit, and neighbours are sometimes of kin, often migrants from the same villages (Mitlin 2014), which enhances the social infrastructure, coordination, and consolidation.

Coordination between neighbours in addressing a common problem, such as an extension of the workshop onto the streets, putting up shared plastic awnings over the street during monsoon, and sharing of lodging responsibilities of seasonal migrant workers, brings social cohesion. In contrast, competitive practices, in this case, trigger contestation, often resulting in feuds over potential customers and discontentment due to each other's everyday practices. These feuds sometimes result in a loss of customers and potentially disrupt each other's businesses (interview: Kumartuli 2018). If one *kumar* does not agree to lower the price of idols, customers would often approach another *kumar*. However, in Kumartuli, certain everyday practices have been collectively legitimised by the residents that might otherwise be perceived as encroachment or illegal in more 'formal' neighbourhoods. As mentioned earlier, streets are spaces of social and cultural exchanges and extensions of the workshop spaces. Figure 4.8 illustrates such organised collectives in the 'in-between' street space of Kumartuli. While there might be dissatisfaction over too much encroachment of streets and using these spaces for everyday activities such as clay-mixing and resting smaller clay-sculpted pieces of idols during the day, in general, most residents accept such practices due to the acknowledgement of lack of space within the workshops. It seems that where resources are scarce, neighbours negotiate and come to a common ground for mitigating everyday struggles with infrastructure in order to come to an amicable solution. They begin to live with the adjustments by extending spaces of the workshops into the streets or, where possible, by shifting the spatial practices beyond the boundaries of the workshops in an attempt to reconfigure the spaces to fit the practices.

The cooperative office of the artists (Figure 4.8, R3–2) plays a vital role in the support network of the local businesses. All the smaller businesses in the neighbourhood and the potters' community are not only part of a more extensive supply chain system that qualifies the place but also part of this cooperative (*Kumartuli Mritshilpa Sanskriti Samiti*) that supports their businesses in times of need and emergency.

> The aim of the cooperative is to help the kumars and the community. We often help them to secure loans. Kumars cannot afford taking loans from mahajans because of the higher interest rates. So, we mediate with the banks to offer them loans at lower interest rates. If the idols remain unsold, kumars suffer huge losses and cannot repay the amount. We then mediate with the bank and try to assist those kumars. . . . We give all kumar members of the cooperative kerosene oil at a subsidised rate, much lower than the market price. You know, for cooking, working at night, for drying the idols using an oil lamp for three months every year.
>
> (Interview: Cooperative representative, Kumartuli 2018)

For example the cooperative runs a free dispensary for the low-paid and seasonal workers in the Kumartuli area. The free weekly eye clinic benefits a lot of elderly artists in the cooperative premises (interview R3). The participant, whose workshop is across the road from the cooperative, feels much attached to the community and relates his success to the support from the cooperative and close friends in the neighbourhood.

> [S]ometimes orders come through friends; if they are too busy, they would refer clients to me, and I would also sometimes do the same. We all try to

help each other. If we move to a new place, we will not know our neighbours; we will not get the support when we need.

(Interview R3)

Another participant feels proud to be part of a close-knit neighbourhood, where he has grown up around friends and cousins and learnt life skills from elders and of kin. However, he feels the neighbourhood is losing its character as people are moving out and seasonal workers replace families of residents, and some workshops are closing down due to the artist's old age and lack of family support. He (R6) is determined to live in the neighbourhood that once belonged to his forefathers and will continue encouraging younger people to do so (interview R6).

Most of the residents are conscious and empathetic of the lack of space in each other's workshops and houses—

[T]hey almost live on the streets because they have lack of space to live in their house; government or some NGO should do something because ours is an important crafting profession. So many foreigners come and film us; I am hopeful that something good will happen to our neighbourhood; after all, we have so much religious and cultural heritage.

(Interview R1)

A feeling of respect for coordination among residents over the shared resources like the narrow streets and communal water taps are reflected in many interviews.

However, a few instances were also drawn where neighbours' actions have triggered arguments and discontent. Participants expressed grievance due to the daily habits of their neighbours. For example abandoned bamboo frames, structures, and raw materials disposed of on the streets cause an infestation of mosquitoes and other pests that spread diseases. The 'relentless' feeding of stray dogs causes them to defecate in the open on the streets, causing odour and health hazards (interview R9). A common concern for most participants was the extensions of neighbours' workshops on the streets through awnings which caused the already narrow streets to feel even narrower. Although extending workshops to the streets result in narrower roads, it also helps foster a close-knit community through constant social interaction and mutual support during the production phase. Sometimes, during the peak season, residents even help out with each other's work.

If my delivery of idol is delayed due to some reason, the calendar will not delay the festival— . . . the idol has no value after the day of the festival. As time is the essence of our profession, we help out each other to make sure we get help when we need it the most. . . . I will not lie to you; we do get a lot of support from our neighbours.

(Interview R6)

In Kumartuli, individual rooms in large houses are home to families of kin (interview R6). The allied industries were able to develop and expand from Kumartuli due to the existing flows and strong social networks.

Will Kumartuli continue to thrive?

Participants in the photo-study belong to different age groups and professions and have different family biographies. All the participants had different approaches to constructing the stories of the places they were attached to, even if these were about the same locations. For example the younger participants, like R8 and R9, had a different meaning to the riverfront than the middle-aged participant R5. Also, the oldest participant, R10, who was sceptical about operating the camera by himself, took me along with him while taking the pictures. He expressed an attachment to more places than others due to his age and lived experiences. Younger participants were generally more concerned about the practicalities of the place: the infrastructure, the services, and the frustration due to a lack of voice in the local decisions. Again, R7, the only female participant, had pictures of her kitchen, which none of the male participants had or was concerned about. They were more concerned about the economic meaning of the workshop-spaces or the transport linkages that affect their businesses. Maybe, roles in the household are still gendered, but due to the lack of participation of women in the photo-study, or rather the absence of women in the front rooms of the workshops during my fieldwork, it is not possible to come to a nuanced understanding on this matter.

A comparison to illustrate the significance of the photo-study can be drawn by studying the contents of photographs that are location-wise similar. Lombard (2013) observed the significant difference between the pictures that the researcher and the residents took. It is often observed that residents of a specific locality 'produce and interpret' images differently than those produced by others. In the context of marginalised groups, the representation and discussion of images become a tool for emphasising the inequality prevalent in such places and are perceived as a tangible outcome of the research process. The contents of the photos differ in terms of the angles of the pictures and the presence of people. In participants' cases, pictures of people they know and interact with daily were very common. However, as a researcher, I tended to take pictures avoiding people or had to ask for consent before clicking, a prerequisite of my research ethics. Figure 4.9 is an example of this. My picture shows the spaces carved out from the streets to accommodate everyday practices but does not include human activities. On the contrary, Figure 4.9, by R5–19, shows the activities, like cooking and interacting with neighbours, that characterise these spaces.

Also, the meanings of the photographs participants took varied largely from those I took. The residents only conceived the sensory perceptions of the alleyways, and I, the researcher, interpreted these spaces as chaotic and complex. One interesting example was that of the *Dhakeshwari* temple [Figure 4.9, R5–12]. The story of the *Dhakeshwari* temple would not have been known to me with only interviews. During my studies, I did not take snaps of the temple. I took a perspective view of the street on which it is located to visualise how the outer streets of the neighbourhood have also integrated as part of the idol-making cluster, completely unaware of the importance of the temple. The original *Dhakeswari* temple, situated in Dhaka, is one of the most sacred Hindu Bengali temples and is often called the national temple of Bangladesh. During the partition of Bengal in 1947, the approximately 800-old deity was moved to Kumartuli, where the *Dhakeswari* temple was

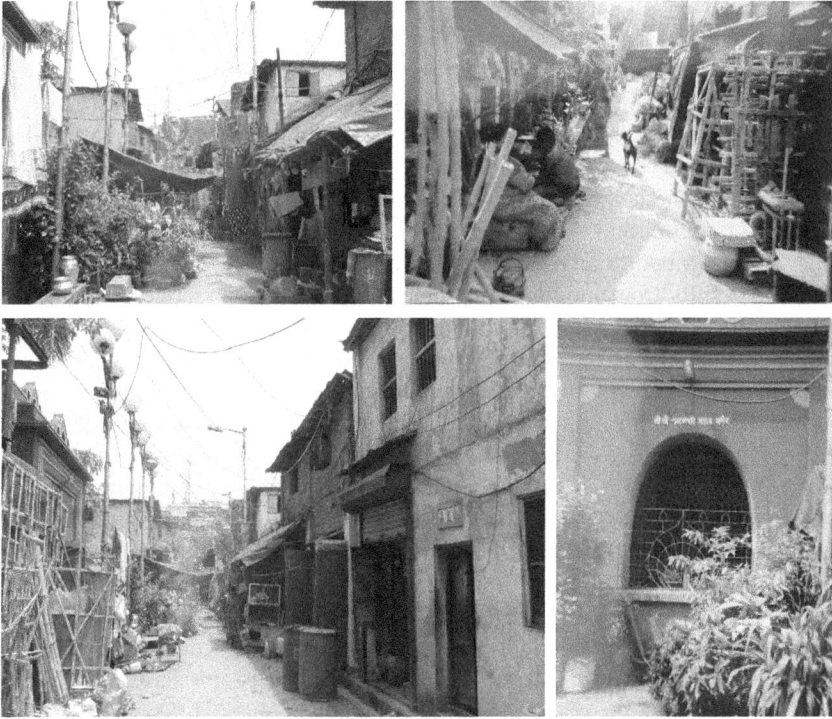

Figure 4.9 Top: streetscape and everyday practices [*source*: author (left), R5–19 (right)]; bottom: images of the street of *Dhakeswari* temple [*source*: author (left, Kumartuli Street overlooking *Dhakeswari* temple on the left) and R5–12 (*Dhakeshwari* temple gate, right)]

rebuilt. The original deity remains in this temple. The fact that the *Dhakeswari* temple was built after Bangladeshi potters settled in Kumartuli unfolds related family biographies.[4] In none of the earlier interviews I had asked about the significance of the places of worship, and people only talked about a support network of ancillary industries related to idol-making. Furthermore, the temple being a part of the dwindling social infrastructure continues to carry on the legacy of emotional and sentimental support and respite to the residents of Kumartuli. The location of the temple is outside the core *Thakurpotti* area, but the current residents have integrated it into the neighbourhood, and people, irrespective of the origins of their ancestral homes, worship at this temple. Religious symbolism and practices in a neighbourhood that makes religious idols for a living might seem unsurprising. However, images produced by the participants show attachment to different temples due to different religious sentiments and emotional bonds. This is an unheard story of the temple and the families that worked hard to build it; is it fair to say then that displacement would disrupt these ritualistic everyday practices of visiting the temple?

The photo-study also raised more specific concerns about particular streets and the residents' struggles. I took pictures to document the built character and infrastructure of Kumartuli. As part of my ethnography-inspired spatial study, I documented how the workshops encroached on the street. However, the pictures taken by the participant, who is regularly affected by the dysfunctional condition of the street, narrates different meanings attached to each picture.

This participatory photo-study unfolds stories of ordinary places narrated by ordinary residents of Kumartuli, but each different story collectively adds to the uniqueness of the neighbourhood. The study brings to light wider implications of social, cultural, and spatial characters that help construct the places of the inner-city neighbourhood of Kumartuli. The construction of these places is complex and results from individual agencies and everyday practices. These ordinary people construct places of unique character every day, which is essential to understand the place as a whole. However, the local government was inclined to build futuristic redevelopment plans to promote tourism that capitalises on the potential of Kumartuli but completely ignored the everyday practices that have evolved in place for over centuries. This participatory research has unfolded nuanced understandings of the neighbourhood's socio-spatial arrangements that produce meanings and attributes to a marginalised community involved in crafts' production, distinct from the generalisations of informality from the global South cities. This chapter, by using multiple and multi-dimensional views to express the meaning of place within the neighbourhood, unfolds the question of attachment, belonging, and aspirations: how older and younger people have differing views, how more famous and the struggling people have more differing views and how the displaced and the 'remainers' have differing views, and also how potters and *shola* (pith) merchants have completely differing ideas of the neighbourhood, the development potentials, and the future. A combination of the researcher's and participatory photostudy findings on how residents of a densely populated *basti* interact with the existing uneven physical infrastructural systems and overcome these precarities contradict mainstream images of an urban *basti*.

Placing the practices within the spaces of informality and creating blurred boundaries of the ordinary private and the public places of the neighbourhood make for a greater understanding of place-making within nexuses of practices and spheres of interwoven activities. Also, this chapter illustrates how these everyday practices lead to the construction of place within the wider framework of the city—the neighbourhood and the supply chain network that contributes to the city's economy as a whole. The practices often overlooked as everyday practices of the residents of such *basti* neighbourhoods, however, play a major part in shaping the urban fabric and characterising the inner-city areas in more ways than producing idols for an annual festival.

Notes

1 For details on *pandal*-construction, crafting, and associated ephemeral structures erected during Durga Puja, see Chattopadhyay, S. (2019). Ephemeral architecture: Toward radical contingency. *The Routledge Companion to Critical Approaches to Contemporary*

Architecture. S. Chattopadhyay and J. White. New York and London, The Routledge: 138–159.

2 The first number on each photograph corresponds to the participant annotated as 'R' meaning the resident, followed by his photograph number. The photographs on each plate, therefore, is numbered as 'R (corresponding number on the table)-(photograph number)'. See 'methodological appendix' for details.

3 Usually self-built, but might also mean that construction techniques and materials are locally sourced and often assembled with semi-skilled workforce and using local knowledge.

4 See Bhattacharya, T. (2007). Tracking the goddess: Religion, community, and identity in the Durga Puja ceremonies of nineteenth-century Calcutta. *Journal of Asian Studies*, Cambridge University Press. **66**: 919–962.

References

Bean, S.S. (2011). The unfired clay sculpture of Bengal in the artscape of modern South Asia. *A Companion to Asian Art and Architecture*. Malden, MA, Wiley Online Library: 604–628.

Bhan, G. (2019). Notes on a Southern urban practice. *Environment and Urbanization*. **31**: 639–654.

Bhattacharya, T. (2007). Tracking the goddess: Religion, community, and identity in the Durga Puja ceremonies of nineteenth-century Calcutta. *Journal of Asian Studies*, Cambridge University Press. **66**: 919–962.

Chakrabarti, D. (2022). Sustaining spaces of idol-crafting and communities of practice: Seasonality, adaptability, and cultural identities in Kumartuli, Kolkata. *The Jugaad Project*. **5**.

Chattopadhyay, S. (2019). Ephemeral architecture: Toward radical contingency. *The Routledge Companion to Critical Approaches to Contemporary Architecture*. S. Chattopadhyay and J. White. New York and London, The Routledge: 138–159.

Friedmann, J. (2010). Place and place-making in cities: A global perspective. *Planning Theory and Practice*. **11**: 149–165.

Ghertner, D.A. (2014). India's urban revolution: Geographies of displacement beyond gentrification. *Environment and Planning A: Economy and Space*. **46**: 1554–1571.

Hannerz, U. (2003). Being there. . . and there. . . and there! Reflections on multi-site ethnography. *Ethnography*. **4**(2): 201–216.

KMDA (2009). Kumartuli Urban Renewal Project.

Lewicka, M. (2011). Place attachment: How far have we come in the last 40 years?. *Journal of Environmental Psychology*. **31**(3): 207–230.

Lombard, M. (2013). Using auto-photography to understand place: Reflections from research in urban informal settlements in Mexico. *Area*. **45**(1), 23–32.

Lombard, M. (2014). Constructing ordinary places: Place-making in urban informal settlements in Mexico. *Progress in Planning*, Elsevier. **94**: 1–53.

Marcus, G.E. (1995). Ethnography in/of the world system: The emergence of multi-sited ethnography. *Annual Review of Anthropology*. **24**(1): 95–117.

Mitlin, D. (2014). Politics, informality and clientelism—Exploring a pro-poor urban politics. *ESID Working Paper No. 34*: 36.

Pink, S. (2012). Situating everyday life: Practices and places. *Situating Everyday Life: Practices and Places*. Los Angeles, CA, SAGE Publications.

Simone, A. (2004). People as infrastructure: Intersecting fragments in Johannesburg. *Public Culture*. **16**(3): 407–429.

5 Complexities

The redevelopment plan

In the previous chapters, I highlighted some aspects of the crumbling built infra-structure and public opinion around it in Kumartuli. This chapter examines the local government's failed attempt (in 2011) to impose a spatial restructuring on the area to 'modernise' it in response to its growing importance as a 'centre' to promote cultural and tourism industries (KMDA 2009). I elaborate on the redevelopment proposal to understand the reasons for the residents' collective resistance to the KMDA plan. It will be revealed through the narrative how with the shift of power from one political party to another, the approach to Kumartuli and the subject of cultural heritage and tourism dynamics have also shifted. While the former left-front government attempted, albeit failed at, a socio-spatial restructuring of the crafting community, the current Trinamool Congress-led government has taken a neoliberal approach to promote the festival of Durga Puja and allied businesses. Finally, the chapter ends by raising questions about the informality and duality faced by the residents due to the land and building ownership, the colonial system of 'Thika Tenancy'[1] in *bastis* of Kolkata, and the age-old rental agreements that pose challenges in building works. This chapter highlights the social, spatial, cultural, and political relations within the idol-making industry and analyses the wider issues of the informal sector in India.

Through the different sections of this chapter, I study the plan's effects and how it forced abrupt changes in the everyday practices of some idol-makers. I also dis-cuss how, after the change in the steady government in 2011, the current policies affect existing practices. Interviews during fieldwork (in 2017 and 2018) and the further round of investigation (deliberative workshop, 2022) highlighted that idol-makers largely believe that Kumartuli is theirs. At the same time, the *shola* mer-chants think that they are equally important resident businesses, revealed in the aftermath of the redevelopment-project-raised questions on the social positions, politics of power, and agency. Subsequently, the implications of the initiation of the plan had repercussions leading to contestation and competition between the resident families and strained relationships with the potential loss of social capital in general. These slow changes in everyday lives in the neighbourhood and the discontent of the residents have been presented through the different sections to

DOI: 10.4324/9781003341222-5

understand the stakeholders' perception towards their immediate surroundings and the places they work in.

Between 2008 and 2011, the State government of West Bengal, through the Kolkata Metropolitan Development Authority (KMDA), proposed a redevelopment plan for the Kumartuli area, with a special focus on the idol-making cluster. This chapter discusses in detail the spatial layout and the proposed physical infrastructure upgradations of the plan, and how it envisioned capitalising on the tourism potential of the neighbourhood. Here, I also discuss the plan and begin to unpack the failed project's implications within the stakeholders and the targeted beneficiaries. Using data published on the KMDA website, drawings, and reports collected from archives at KMDA, I studied the proposed plan and interviewed planning officials and current and ex-local political representatives.

The KMDA's Kumartuli idol-making cluster redevelopment project laid out plans for restructuring on a spatial scale and did not consider the complexities of land ownership at the planning phase. The project's stated aims were to 'understand' the 'social and physical infrastructure and cultural heritage', which would guide the 'urban renewal' process for actual intervention through the 'adoption of regulatory and institutional mechanisms' (KMDA 2009). There was also a statement about the possibility of 'regeneration of the economic pursuits' through 'landuse planning and innovative zoning' to support the existing economic activities and generate more revenues. A combination of slum redevelopment and an upgradation scheme was proposed, where redevelopment implies the demolition and rebuilding of dwelling units based on building plans drawn up by the KMDA, and upgradation means improvement of basic infrastructure and services like the roads, electricity, and drainage. The vision, however, in addition to the stated aims, was to develop Kumartuli as a potential international tourist destination of Kolkata (interview: Kumartuli 2017).

In the statement, the plan aimed at an overall regeneration of the Kumartuli neighbourhood. The project would only benefit those involved in economic activities related to idol-making, including practices such as *shola* (pith) carving and ornaments and garments shops, that is the interwoven communities of practices relating to idol-making based in the Kumartuli neighbourhood. However, the plan failed to consider the other residents who are not involved in idol-making practices or related directly. Although the aim stated the scope for adopting regulatory and institutional mechanisms for the social and physical infrastructures, the proposal was to build in situ slum redevelopment with temporary relocation and 'slum-networking'.[2] This means the materialisation of the plan would only involve the demolition and rebuilding of workshops, residences, and building-related services under the slum-networking programme. The slum networking proposal may be debated, as Kumartuli, within the inner wards of Kolkata, is already within the networked (piped) sanitation service area. However, it could be pointed out here that due to the lack of toilets currently in the residences of Kumartuli as a whole, this idea was promoted to build toilets connected to the water supply and sanitation as part of the networking scheme. However, as published in the KMDA report, the proposed plan does not detail the provisions of the regulatory and institutional mechanisms (KMDA 2009).

The proposed 'urban renewal' project in Kumartuli covered an area of 3.73 acres and 522 beneficiaries (households) over two to three years. During these two years, the affected people would be relocated temporarily to a nearby warehouse location, about a ten-minute walk away from the neighbourhood. The warehouse was remodelled for temporary relocations with dwelling units on the first and second floors, and the ground floors were reserved for the workshop and storage areas. The phase-wise relocation scheme meant that there would be cycles of beneficiaries moving in and out of the warehouses as and when their residences would be affected. In other words, they would only be housed for a short span of time, during which their respective residence-workshops would be demolished and rebuilt, according to the plans.

The plan relied heavily on survey records and proposed allocating as much square-footage as a household occupied in Kumartuli previously, as recorded by the private surveying agency hired by the KMDA or as recorded in earlier building plans. There were also suggestions to the residents that families would be entitled to as much space in the new development as they had in their previous residences. However, the drawings laid out were not in compliance with the survey or earlier suggestions. A consistent layout of family units of 25 square metres (269 square feet), including a multipurpose room, a bedroom, a small kitchen, a bathroom, and a veranda, based on minimum housing recommendations of an earlier central government-recommended slum-redevelopment scheme called the Rajiv Awas Yojana (RAY),[3] was proposed. The original much-debated RAY scheme aimed to provide housing for the homeless and impoverished urban groups and envision 'slum free cities'.[4] Whether or not the space allocation would be justified to the needs and requirements to fit the practices of Kumartuli remains a question. The new buildings would be three- or four-storied, and the ground floors would be for workshop and storage spaces (KMDA 2009). The plans were not only ambitious and apparently beyond feasible within the limited territorial space available for accommodating the existing practices but would also have meant that some families would have to compromise about the existing built-up spaces while financing their new homes. Newspaper reports suggest continued discontent and debate between the planning agency and local community representatives (artists' cooperative) about space allocation, temporary relocation, and the project's timescale.

In addition to the development of new buildings for accommodating workshops and residences, there would have been provisions for a small (four-bedded) health care centre, a four-storied 60-bedded dormitory (supposedly for seasonal migrant workers and supporting the training facilities) with pay and use toilet facilities, training centre, exhibition-cum sales counter, 'international standard pay-and use toilet', a green park, parking facilities for hand-drawn carts and rickshaws, a transformer, wide roads with pavements, drainage and underground water storage facilities, and also garbage disposal points (KMDA 2009). The proposals aspired to house seasonal migrant workers and build training centres to promote the dissemination of idol-making skills. The plan shows a lack of nuanced understanding by the KMDA planners and the government of the needs of the residents of Kumartuli and their everyday practices. I present the key setbacks of the KMDA plan and structure them according to three key aspects: provision of space and practicalities, consideration of material practices, and the incremental housing requirements in constrained informal spaces.

First, the spatial restructuring proposal and subsequent top-down land use planning proposal caused discontent among residents. The materialisation of the futuristic vision of the 'tourist-hub' of the government would only be realised through a densely built medium-rise apartment-cum workshop block constructed with 'cookie-cutter' building plans with low-quality bricks and mortar for housing the 'poor' to reduce project costs. There were no details of the location of such new facilities and buildings as there were already implications of space constraints in housing all beneficiaries. Also, there was no consideration as to how and who would run and fund these proposed facilities. The plan that set out to incorporate the government's aspirations to tap into the neighbourhood's tourism potential only proposed building a marketable solution for the largely apparent physical infrastructure issues like garbage disposal, narrow alleys, electricity, sanitation, and drainage.

The plan may not have consciously considered the everyday lives of residents, including the 'street' as an essential space for social interaction. As presented in previous chapters, most residents of Kumartuli consider the street as an extended part of the domestic space and associate the street with performing productive functions. As (Dovey 2013) highlights, most formalisation plans standardise the private spaces that might not always have as much ready access to the street as before, disrupting the flexibility and productive functions of the space. The residents scrapped the plans because not only these were not reflective of their needs and requirements, but these also failed to address the coordinating and cooperative aspects of the apparently thriving practices making up the majority of the idol-making industry that would continue to generate economy if provided with a boost.

Second, the report provided no rationale for the proposed residential designs and the commercial and infrastructure facilities. Nor were the design requirements stated, so there is no way to know whether this plan met the neighbourhood's requirements as a whole. Here, the spatial mapping studies reflecting the everyday practices within houses and streets of the neighbourhood (detailed in Chapters 3 and 6) help to analyse these plans. For example, the proposed plans do not include the relationship between the open production spaces and the apparently 'inaccessible' mezzanine floors, which are essential spaces for storage and related activities of idol-making practices. Also, even if the houses are vertically arranged, it would always be necessary for the artists to have a ground floor workshop-cum storage space for the distribution and production of idols. Upper storeys could only be for residential purposes. Even then, everyday informal interactions between neighbours, passers-by, and possible clientele would only be restricted during workshop-opening hours. Wide roads and pavements could potentially disrupt everyday interactions between neighbours. However, these were certain subtle ways, unlike the proposal of an international standard toilet and exhibition-cum-sales centre, through which it was indicated that the aim of the government above the basic provision was to market Kumartuli as a cultural and heritage tourism hub on the basis of the planners' and politicians' perception of westernised standards of modernity.

As presented in the literature review, idol-crafting using traditional practices of clay and straw contradicts modern living views. Interviewees both in Kumartuli and the planners and politicians have expressed a general discourse of associating these practices as muddy, uncleanly, and dated. While idol-makers have expressed the need for having the spatial–material configurations of the built form, the plans suggest that the proposed development would have been far from it.

Third, among other criticism of the plan, there is no mention of the scope of future expansion and incremental mechanisms. Was it indicated that the families that now reside and work from Kumartuli would permanently be involved in the practice and pursuit of idol-making or be constrained by the allocated 25 square metre residential units? Moreover, the previous government strongly envisaged the potential of the idol-making quarter and planned to develop tourism based on the cultural and heritage assets of the place. This idea is also reciprocated among the few famous artists who currently do not reside in Kumartuli anymore (interview: Kumartuli 2017, 2018). The budget for the project was in the region of 26.80 crore INR in 2011 (approximately 28.6 million GBP), out of which the share of the central government's contribution would be 35 per cent, and the state government would provide for 45 per cent. Furthermore, the rest would have to be paid by the beneficiaries, and the KMDA would arrange for loans to help them. Moreover, the plans were generalised to provide minimum living and working spaces on the basis of certain national housing prototypes. It lacked the scope for extension or the provision for future expansion of workshop spaces.

The idea is also questionable; although the housing was affordable, was it discussed and agreed upon with a general consensus? Were the voices of the residents heard on this matter? My questions were faced with blank responses from planners, and the idol-makers, too, seemed unaware of planning consultation procedures (interview: Planners, Kolkata 2018). The residents of Kumartuli have homes and workshops they rent or have owned for a long time; hence, building new homes for them would potentially have an associated cost. Moreover, whether the residents were notified about these questions remains unclear from the stakeholder interviews. Either they were unaware of the answers or chose to remain quiet on the matter and only express discontent about the failed initiative that raised their hopes of better living.

While most interviewees in Kumartuli did not say much, the planning officials involved in the project expressed frustration with its failure (interviews: Planners, Kolkata 2018). Planning and executing committee officials at KMDA recalled the incidents in the interviews. They cited two reasons for the project to come to a standstill. First, the number of beneficiaries increased by at least 7–8 per cent over a few weeks; hence, the resources were exhausted. Overnight, one of them said, *'people started converting their two-storied structures to three or four-storied ones and bringing in new tenants'* who could well be their relatives from the village (interviews: Planners, Kolkata 2018). Second, this was the time when the shift of political power was happening in West Bengal, and the beneficiaries supported by the two parties began an internal conflict over the spaces provided through the proposal and the relocation scheme during the redevelopment of the area. The conflict arose over the allocated spaces between the idol-makers and the *shola* merchants. As per the KMDA plan, the idol-makers were allocated 20-foot-high storage space and workshops to allow enough headroom for their idols, whereas the *shola* merchants and the jewellers were allocated rooms according to the minimum habitable building standards of the Indian National Building Code. Hence, the conflict was over not only the allocated floor area but also the height of buildings and the number of stakeholders and beneficiaries on the list.

When the space issue was amicably temporarily sorted out with the intervention of the public representatives and an open discussion between the two parties, rehabilitation for the first phase of work started. In the meantime, the issue of land ownership came to the forefront. The dated and precarious land and building ownership deeds were one of the major reasons that the KMDA rehabilitation project for Kumartuli failed. There are a few pockets of privately owned lands in-between the Thika Tenancy plots, and both groups of stakeholders could not be provided with the same benefits due to the complexity of the rental agreements and ownership statuses. According to the plan, the private landowner would only get the number of rooms they and their immediate families occupied. All existing long-term tenants would also get the same benefits, and the new tenants, brought in to occupy more rooms in extended houses, would also get similar benefits, which to the building owners seemed unfair.

> The people, who converted their houses overnight, felt cheated upon and fooled; hence they approached their local party leaders again and collectively disrupted the demolition work. The few families, who were temporarily shifted, remain in that warehouse and continue to use both.
>
> (Interview: Planners, Kolkata 2017)

The authorities could not come to a decision on the stake of the beneficiaries and discarded the plan altogether (interview: Political representative, Kumartuli 2017). The few people who were rehabilitated were using the workshop places provided during the peak season for storage. In the off-season, the spaces are mostly empty, damp, and dark, attracting antisocial activities (interview: Political representative, Kumartuli 2017). In the aftermath, the warehouse has been conveniently renamed 'Kumartuli 2', associating it with the idol-making industry practices while retaining the brand name to a potential customer. However, whether they wanted to retain the same brand value for the two places by this renaming and continue to thrive from both remain unknown.

Reaction and resistance to the KMDA plan

The reasons for the resistance to the KMDA plan were complex. The media covered the resistance, and the news was all over the city (Datta Roy 2008; Das 2009). Most of the reactions recorded during the interviews to the plan pointed to the political duality faced before the 2011 state assembly elections and the subsequent power shift. Others have implied the precarious ownership conditions in the Kumartuli neighbourhood and the issues relating to the allocation of spaces. None of the interviewees were particularly happy to talk about the aftermath of the haphazard implementation of the plan, nor were they interested in reflecting on the details of the failed plan other than mentioning it briefly. Also, during the interviews, the respondents requested anonymity and not to record sensitive information. They generalised the reasons for failure and blamed the local leaders for causing the precarious situation. The shift in political leadership and the change in the long-standing government resulted in political factions within the neighbourhood. The locally elected representatives in favour of the Trinamool Congress Party (TMC) (currently in power) resisted the conceived plans during the previous Left Front government.

It must be stated here that in West Bengal, grassroots level organisations of political parties (locally called party offices, that became more common during the Left Front Government, c. 1977–2011) play an important role in the neighbourhood structures and influence politics at the local level while also networking with other nearby neighbourhoods and at the municipal level (Roy 2003). These party offices were controlled by middlemen who liaise with the councillors (elected municipal representatives) while also maintaining clientelism[5] at the neighbourhood level. Opposition parties also have a similar network of party offices and local intermediaries. There was a perceived shift of power among the public through the local party offices before the shift actually happened due to frustrations with the then government (interview: Kumartuli 2018). The local leaders and their middlemen knew of this recent power relation and capitalised on it to further their political aspirations. Collective action in the resistance to the plan due to this newfound strength in the opposition party (TMC) was triggered by a majority in the 2009 municipal election, who eventually came to power in the State Assembly in 2011. Furthermore, it was around this time that the plan was discarded altogether.

The interviewees did not express their affiliation with or support for political parties and only blamed the middlemen as the force to lead the collective action. Some said the middlemen, locally designated as the '*para*' goons, always remain the same even if the parties in power change; the goons only changed sides. None of my interviewees were actively involved in the resistance; however, some views were expressed through the approach to their discussion on the topic of redevelopment. Although most interviewees seemed unaware of the details of the proposed developments and how they would benefit from it, those who were in support of the entire redevelopment project patiently waited for the project to be successfully completed; some moved to the warehouses, others hoped to have a newly built workshop and residence. While some benefitted from two workshop spaces, others were made to anticipate and finally give up on the hope of receiving any benefit from the plan. Those who supported the resistance were perhaps hoping that a future plan initiated by the new government would be more democratic, reflect on the needs, and be sympathetic to the practices therein. Although it had been several years since 2011, the interviews with the political leaders at local and regional levels did not suggest any further redevelopment plans (interview: political representatives, Kumartuli 2017). However, a strong support for tapping the tourism potential of Kumartuli was perceived through the 'cultural industries' policies of the current government and resonated in the interviews with politicians. Perhaps, these views were echoed through the nomination and, ultimately, recognition of Kolkata's Durga Puja in UNESCO's ICH list.

The reactions differed among different groups of artists, while most resisted and opposed the proposed plan. Some younger artists accepted the redevelopment plans as they thought it was '*insurance for a successful future*' (interview: younger artist, Kumartuli 2018). They accepted the temporary relocation proposal and moved to Kumartuli 2 [Figure 5.1]. In the summer of 2018, they continued to live and work in Kumartuli 2 and were slowly rebuilding their demolished workshop by their own means on their old plots at Kumartuli. Their practices have been disrupted in the last few years, and businesses have been slow. However, their neighbours did not show cooperation, and potential customers were lost to contesting 'neighbours' (interview: Kumartuli 2018). Apparently, what looked like a gain on the part of these families was a loss of social capital and disruptive in terms of their steady business and clientele in Kumartuli. In another instance, the interviewees looked forward to a better workshop and hoped their workshop would be demolished and rebuilt (interviews: Kumartuli 2017, 2018). However, they are now discontent about the 'unfair inequality' imposed by the government plan: some people within the community enjoy two workshop spaces while some others had

Figure 5.1 Pictures of the temporary living and working spaces of Kumartuli 2, provided as a relocation, photographed in November 2017

Source: author

to rebuild the demolished workshops themselves, and a sizeable number of artists still have to continue business by their own means like before. The failed plan may have resulted in strife within a community rich in social capital. The discontent is most among the last group of people as they felt that they received no government support. However, a common thread in all the interviews and participatory photography was the disruption of everyday practices due to the redevelopment project. Place attachment was the key in all such discussions on redevelopment and subsequent rehousing initiative talks; some said,

> [T]he place would not be the same anymore'(highlighting the uniqueness of Kumartuli), 'I now know my neighbours, but on resettlement, would I know (be familiar with or get on well with) them (?), and I would have to adjust to new neighbours (?)', or 'this house belonged to my father and grandfather; hence I refuse to leave this.
>
> (Interviews: Kumartuli 2018)

As presented through the photo-study in the previous chapter, residents feel attached to neighbours and largely collectively coordinate their efforts. One of the reasons for resistance to the KMDA plan reflected through the interviews was that people would not have an opinion about their new neighbours. This shows that although the plan mostly promised to rehouse all the families in their respective spaces previously occupied by them, people were unsure and sceptical about the idea. Also, planners mentioned that while they tried their best to comply, in some cases, it would not have been possible to do so (interviews: Planners, Kolkata 2017). The plan showed a lack of clarity and future provision to the stakeholders. Therefore, most people remained sceptical and believed they were not allowed to '*choose their neighbours*' (interview: Kumartuli 2018) and would face difficulties adjusting to their new neighbours if relocated elsewhere.

A sizeable number of interviews, mainly from renowned artists, reflected how the Kumartuli neighbourhood, as the 'brand', could be an international tourist hub. They concurred with the idea of showcasing the crafts to international tourists and 'clean' roads and riverfronts. However, none of them implied that the KMDA plan was the answer to that; rather, they wanted good governance and better efforts from the municipal corporation in providing services. Either the plan was not well-communicated with the residents, or the plan perhaps failed to address the cultural and material practices that make the Kumartuli neighbourhood unique. I reflect on Hosagrahar (Hosagrahar 2005) and argue that blind and sporadic replications of western modernity[6] in contrast to indigenous modernity result in splintering urban pockets and uneven development in the Southern cities. The latter idea manifested through the 'world-class' or 'international' standards would be an ill-conceived notion of replicating the western standards with complete disregard for indigenous idol-making practices. There was a lack of place-based appreciation in the KMDA plan, which failed to identify the existing resources, road and river networks, social and cultural capital, buildings, and uniquely set-up shop frontages. The government's oblivious stance to the place-based practices of Kumartuli and the failure to acknowledge the strengths of the 'place' itself as part of the planning process pose questions of governance. Contrary to the public sentiments of Kumartuli's unique-ness, a local politician even referred to idol-making practices as '*dirty, muddy and hazardous*', and the neighbourhood needed a facelift and clean-up from the resi-dents to appeal to tourists. This view was expressed with complete disregard that neighbourhood cleanliness and garbage clearance are part of the services provided by the local municipal authorities (interview: Politician, Kumartuli 2017). Never-theless, residents suggest that garbage collection and road cleaning are infrequent and inadequate.

The reactions of shopkeepers selling the jewellery and outfits differed from the artists. A *shola* and jewellery shopkeeper voiced his requirements for the upgrada-tion of the infrastructure to enhance the 'quality of the place' in Kumartuli's inner alleys. His shop is situated in the internal alleys, the older part of Kumartuli—when there were fewer idol-makers, even the jewellery stores were fewer and concen-trated close to the idol-makers' quarter in that area (interview: Kumartuli 2018). Although he did not have a clear view of the sort of upgradation he actually wanted for Kumartuli, he wanted the project to go on. He blamed the result of the assem-bly election for the scrapping of the proposed development. He would have been temporarily shifted at the beginning of the work and eventually relocated upon project completion. However, the project stopped halfway through the demolition process; work was held up, and when it finally stopped, he had to rebuild a section of his demolished shop and continued business from a makeshift setup in 2018. His major discontent was that he would have received a similar set-up in the relo-cation space as all other families would receive. This, according to him, would have disrupted his business, and his clientele would not know of his relocation. He would have only accepted to move if all his competing businesses had moved simultaneously. Notably, despite having close working relations with idol-makers' cooperative committees, they could not receive a favourable deal, but they had to rebuild the demolition-affected part of the shop and continued business through this appropriated position.

Nevertheless, similar ideas of resistance were reciprocated among other shop and small business owners; they raised concerns regarding the lack of dialogue between the planning agency and the residents about the spatial set-up of the relo-cation programme. The residents only saw a preliminary site plan of the project with the proposed new blocks displayed in the neighbourhood and a list of names of beneficiaries. Apart from a few words about the general failure of the KMDA project, none of the interviewees raised questions about how the plan failed to incorporate everyday practices or commented on their needs. According to the interviews, there was no consultation or participation from the residents in the pro-posed plan. Also, a consultation early in the process might have brought to light the duality in the *basti*'s ownership pattern, resulting in a different approach to the problem. In a way, the lack of public engagement in the planning highlights the perception of the traditional idol-making practices as 'dated' and 'hazardous' on the government's part, leading to stigmatisation and constant marginalisation of the informal settlements.

Over six years of research and fieldwork carried out in Kumartuli, I identified a number of spatial and infrastructural concerns and grievances within the community. I organised a final round of investigation for completing this book in the form of a deliberative workshop in February 2022. This small group deliberation was intended to collectively reflect on these common issues and co-produce recommendations for the much-needed municipal services and infrastructure conditions in Kumartuli. The deliberative workshop discussed infrastructure, development, and inclusive planning with various experts and stakeholders. Further, it also considered how can the precarious conditions of living and working in these slum neighbourhoods be improved with their knowledge, expertise, and experience. Community representatives from Kumartuli attended the workshop, as did Kolkata-based planners, architects, academic researchers, engineers, and policymakers and research students. The municipal engineers agreed that special requirements for neighbourhoods with additional productive and commercial functions require additional services and infrastructure. Hence, they decided to investigate and report this. A similar response was received for solid waste management. Engineers suggested that community representatives voice their concerns to the responsible department regarding their difficulties. Provision of more carts and more daily cleaning services are possible in this case too. These were discussions of very apparent and striking everyday struggles and negotiations of residents with infrastructure, and all present stakeholders thought that improving working conditions and developing policies for better services and facilities that would aid in increased production and ensure the continuation of Kumartuli's heritage of idol-crafting and allied crafts should be a top priority for all concerned parties. Despite its small scale, results from this workshop suggest that it is possible to unpack community needs and incorporate these into planning and governance mechanisms through participatory approaches.

Tenure and ownership: realities

The issues of ownership raised during the redevelopment project have always hindered any building and extension works in Kumartuli. The fact that any addition and alteration are not possible without the agreement of the owners of both the plot and the building causes an obstacle to any new construction. During his interview, a local elected representative (interview: Political leader, Kumartuli 2017) spoke about the underlying complexity of the land and building ownership in Kumartuli and the neighbouring *basti* areas in Kolkata's inner wards. Kumartuli's history as a settlement favoured by the local landed gentry is a major reason for the ownership issues. According to him, the land belongs to one person, and the building is owned by someone else. The building owner had 'in the distant past' sub-let his house on a room basis to different people over time, and these tenancies, mostly based on verbal contracts, have lasted longer than the owners' lifetimes. Some tenancy contracts span over generations of owners and tenants. The complexity of land and building ownership arose from dated tenancy legislation that existed in colonial Calcutta from the early nineteenth century. This system was due to rapid population influx through migration, industrialisation, and uneven socio-spatial growth. The urban poor, mainly the migrants, settled in housing as part of a 'unique three-tier tenancy structure' consisting of the landlord (the owner of the plot), the intermediary Thika tenant (the owner of the hut or cowshed or the temporary structure), and finally the tenant (occupant of the hut or part of it). In 1948, to reserve and secure the rights of the Thika tenants and govern the landlords' rights and liabilities, a 'Thika Tenancy Act' was passed with further amendments in 1952 and 1981. Different rental agreements existed between the landlord and the Thika tenant and between the Thika tenant and their sub-tenants. The building legislation is not very strict, and the local government permits building extensions without formal planning. Also, the relaxation of land and building taxes by the KMC affects people's choices of building materials and the type of finished surfaces like walls and floors used in the building. Lower taxes are levied for building materials classified as 'kutcha' (temporary), for example.

There is ambiguity and deregulation in property planning across cities of India, while the state remains empowered to land acquisition through eminent domain and subsequent change of uses through private developers (Roy 2009). However, age-old rental agreements (mostly verbal and lasting through generations) that reserve the tenants' rights to the occupied spaces (mostly rooms and shared courtyards), construction, or partial rebuilding of houses is mostly impossible without agreement between all occupants. Informal land and building tenures are not restricted only to the poor but also even the affluent middle classes are residents of urban slums (AlSayyad and Roy 2003). In Kumartuli, as much as the residents want to live and work in better houses and have better roads for the smooth running of their business, they would also need to finance these construction works. In most cases, negotiating a fair share of space in the reconstructed buildings and expenses to be shared mutually has proved futile due to the varying resources of the occupant and stakeholder families. Residents, even the affluent, choose to hold on to the cramped spaces in Kumartuli *basti* not only for the inner-city location and the real-estate potential but also for continuing business from the branded periphery of the neighbourhood. Many residents continue to live in cramped conditions or move elsewhere. While a number of *kumars* are unable to afford real estate within the familiar geographies of the inner-city area of Kolkata (interview: Kumartuli 2018), some affluent families have been able to purchase a residence in nearby newly built apartments by private developers (Figure 5.2). The KMC allows for lenient building regulations within registered slums with scope for maximising the allowable FAR (Floor Area Ratio)[7] and relaxation on the maximum plot-covered area. This relaxation of regulation in notified *bastis* allows residents to extend or build houses covering a larger built area on a relatively smaller plot. Some residents, although Thika tenants, have managed to negotiate with their landlords and collectively constructed apartments within pockets of the slum using the regulation relaxation and have accommodated all the long-term tenants and stakeholders within the building.

Explaining the hopelessness of the situation of their immediate environment and the lack of services provided by the government, the respondents in Kumartuli

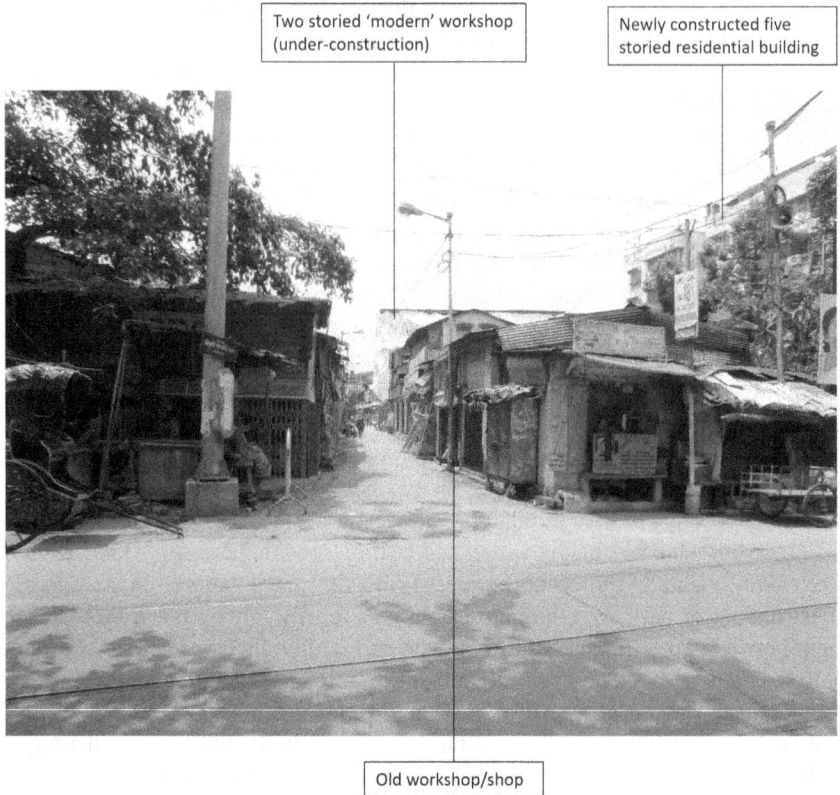

Two storied 'modern' workshop (under-construction)

Newly constructed five storied residential building

Old workshop/shop

Figure 5.2 Mix of building typologies in Kumartuli: view of a street entering Kumartuli's inner areas; while most buildings are self-built, a few buildings are beginning to be constructed out of contemporary materials, photographed in April 2018

Source: author

expressed deep emotions reflected in their interviews. A local resident, explaining the situation, said,

> This place could be developed instantly (through eminent domain). There are few government rules, norms and restrictions, that is why nothing can be done here.
>
> (Interview: Kumartuli 2018)

Although this is a statement of frustration, it raises questions on the perception of the state's power and the residents of a seemingly lower caste group's rights to their neighbourhood and the wider city. While the current Trinamool Congress party in power had assisted in the organisation of the collective resistance to the earlier redevelopment plan as political opposition to the then government, they have since not commented on any redevelopment or infrastructural upgradation plans. However, subtly and incrementally, the present government has promoted the 'world-class' tourism agendas through the Durga Puja festival and the nomination (in 2019) and inscription (in 2021) of the same for UNESCO's Intangible Cultural Heritage List of Humanity.

The stigmatisation and marginalisation in informal settlements are projected with ideas of chaos, disorder, and in this case, the local councillor's notion of lack of 'cleanliness' (interview: Politician, Kumartuli 2017), whereas lack of infrastructure and services is a deficit in the part of city governance. While the Thika tenancy and other rental agreements pose problems to rebuilding, the government's attempt to modernise and redevelop the area to conform to western standards by alienating the practices marginalises the neighbourhood. In doing so, the local government has stigmatised the informal tenure and practices within a thriving informal economy to the point of a general discourse of restricting the use and storage of mud, clay, and waste from the production spaces within the streets of Kumartuli (interview: Kumartuli 2018).

Informality is often viewed as the 'unplannable' or rather as a state of exception from the formal order of urbanisation (Roy 2016a). Global and local economic and political forces have shown to have mostly negative impacts on the marginalisation of informality, both economically and spatially. Urban planning interventions like 'upgradation' or 'redevelopment' projects in informal settlements of the global South have shown to have major economic challenges to the urban poor, who tend to shift spatially as a result (Rao 2018). These shiftings are sometimes triggered by state-led evictions or, in other cases, are simply the effects of enforcing a 'formality' through planning measures and upper- and middle-class actions (McFarlane 2012; Lombard 2014; Banks et al. 2020). Even if the planners would have negotiated an amicable solution among the stakeholders by preserving the rights of each, the result of the KMDA-led planning might have been a further marginalisation of the residents of Kumartuli. People would have been forcibly displaced; practices already viewed as 'dirty and messy' by elected representatives might have resulted in organically shifting elsewhere. Hence, knowledge about practices is essential to build on the theories of informality and understand seemingly unseen spaces of productivity through the residents' narratives and place-making in their immediate built environment.

Informality in the heritage

This section summarises the manifold informal modes of the idol-making industry and the challenges in incorporating them into the mainstream economy. The pertinent question raised here is whether the policy frameworks in place address the issue of an informal (largely unorganised) industry located within a precarious quasi-legal framework of land and building ownership of an urban *basti*. Does the policy framework consider the residents' rights involved in a traditional practice in a transforming mixed-use *basti* neighbourhood constantly facing tourism policies and place-based real estate pressure? Or does the physical and material infrastructure in place barely support the idol-making practice due to the informal tenure of the neighbourhood? As Banks et al. suggest, '*understanding informality requires a differentiated analysis of actors operating within and across domains*' (Banks et al. 2020). The observation encompassing the exclusion of informal groups based on their ways of living, housing, and access and the exploitation of the informal domain by seemingly more powerful groups for maximising profit to avoid taxes and regulation requires further investigation.

The argument for a standard dichotomy of formal and informal has been rejected in favour of the suggestion that '*informality is not a separate sector, but rather a series of transactions that connects different economies and spaces to one another*' (Roy 2005). Also, as presented in the earlier sections, informality is not linked to poverty only but to a number of complexities and multiple marginalising factors. Like many other older cities, Kolkata's inner wards are also character-ised by a dense network of informal sector activities that make up a large part of the economy but are mainly governed by informal modes of exchange. About 75 per cent of India's workforce is estimated to be part of the informal or unorgan-ised employment sector (Chattaraj 2016). The informal sector economy consists of small-scale enterprises; businesses run from home, including artisans and craft producers and their employees who lack social security and legal protections (Unni and Rani 2003). However, recent literature has contested such views (Chattaraj 2016). It is perhaps impossible to strictly negotiate a formal and informal boundary in idol-making practices; it is rather blurred and often overlapping to accommodate interwoven practices.

In Kumartuli, the transactions are also predominantly made by age-old business techniques and norms from the past. Starting from the wage structures to prices of idols, most transactions in the idol-making industry are based on verbal con-tracts and conventional practices, which are constantly contested. Like elements of materiality and competencies in the idol-making practice, the contracts with seasonal workers too are mostly controlled by generational conventions and norms that are slowly changing due to consumer pressure. The absence of a standard-ised wage structure or skills training programmes in technical competencies sug-gests an informal sector economy. However, the idol-making industry has provided employment and business opportunities to generations of potters for centuries. Informality is in every aspect of the industry—the economics and financing of the idols, including the way it operates, as well as the skill development and legacies of the practices. Also, the built environment poses different challenges within the informal settlement domains, such as the land tenure and building ownership issues of the production spaces.

Many high-end Durga Puja celebrations involve larger transactions. Even in these higher-budget celebrations, artists employ seasonal workers on daily wages and verbal contracts. What parts of the transactions are recorded and what slips through the cracks of an informal sector economy are not precisely recorded. Idol-makers and other artists (for *pandal* construction and lighting) are paid on a lump sum basis by Durga Puja committee organisers, and, farther down the chain, all transactions are on lump-sum contracts starting from the smaller contractors to

seasonal workers and suppliers of raw materials. The interviews revealed that seasonal migrant workers and idol-makers with smaller businesses do not fall into the
formal income tax-paying brackets and remain in the small enterprise category,
often unaccounted for in the economy. Also, from the interviews during the fieldwork and other secondary sources, it was impossible to determine either the abso-
lute size of the industry or the number of people employed in it and the number of
seasonal migrant workers involved. Idol-makers, during the interviews, were not
specific about the number of idols they prepared each year, and the prices of idols
are always negotiable and competitive. Idols are not always made to order; larger
and pricier ones are, but smaller idols are prepared for ready sale, and some are
never sold (interviews: Kumartuli 2017, 2018). Hence, the scale of operations of
the industry is an estimated value. In 2019, a British Council report mapped the
economic value of the Durga Puja festival collectively as Rs 32,377 crore INR
(approximately 3.29 billion GBP), where 260 crores (approximately 25.5 million
GBP) (0.8 per cent of the total) was estimated worth of the idol-making pursuit
(British Council 2019).

However, the number of Durga Puja celebrations each year in Kolkata and surrounding areas is recorded from the number of permits issued and the number of
immersions. While there have been attempts to redevelop the neighbourhood and
exploit the tourism potential of festivals, a policy intervention or reorganising of
the idol-making industry has not been on the frontline despite the current large-
scale economic turnovers and global flows. The argument that connects the section
is that postcolonial city governance operates in a way that continuously marginalises the practices based in urban slums (Roy 2016b), including such growing
industries as idol-making due to the practices and socio-spatial character of the
neighbourhood. The question here is, what could be a possible way to incorporate
these industries into the policy framework? The disruption and the fears that had
triggered due to the KMDA plan have paved the way for an organic displacement
and a need-based restructuring of the spaces within Kumartuli. While the redevelopment of the idol-making cluster seems over-ambitious and failed to take off due
to ambiguity of the legal and quasi-legal land and building regulations, perhaps an
infrastructure and service intervention to improve the networks and facilities in the
neighbourhood would have been well-received by the residents to accommodate
the growing industry (Simone 2004).

Although collective actions and a network of the support of neighbours and suppliers govern place-based activities and everyday practices, the individual agency
plays a major role in future decisions and the provision of living and working in
Kumartuli. Everyday practices must take a higher position in the decision-making
process for the future of the residents of Kumartuli. Whether or not the resistance to
the KMDA plan played a role in the recent government tourism and cultural policies
is a question that remains to be answered here. Also, whether or not the residents'
rights and their practices in Kumartuli would be reserved in the future, where Durga
Puja would be showcased as a world heritage festival, remains to be understood.

Notes

1 Thika Tenancy Act, 1981:

> *Any person who holds, whether under a written lease or otherwise, land under another person, and is but for a special contract would be liable to pay rent, at a monthly or at any other periodical rate, for that land to that another person and has erected any structure on such land for a residential, manufacturing or business purpose and includes the successor in interest of such person, but does not include a person:- (a) who holds such land under that another person in perpetuity; or (b) who holds such land under that another person under a registered lease, in which the duration of the lease is expressly stated to be for a period of not less than twelve years; or (c) who holds such land under that another person and uses or occupies such land as a khatal.*

2 It is a citywide, 'community-based' sanitation system and 'environmental improvement' programme. It seeks to upgrade the infrastructure of a whole city using the network of slum settlements as a starting point. The result has been a dramatic improvement in the city infrastructure, with a piped sanitation system, clean rivers, and a much-improved road network. This has all been achieved at a fraction of the cost of conventional approaches. This approach was pioneered in the Indian city of Indore and has been successfully adopted in other cities.

3 RAY—Rajiv Awas Yojana is a 'Housing for the Poor' scheme by the Government of India's Ministry of Housing and Poverty Alleviation, as part of the 'slum-free India' vision, implemented between 2011 and 2013 through local government projects. This scheme aimed at providing shelter and access to social amenities in intervened slums, and beneficiaries would have access to credits for building houses following the RAY guidelines.

4 See [Bhan, G. (2017). From the basti to the 'house': Socio-spatial readings of housing policy in India. *Current Sociology*. **65**: 587–602. Bardhan, R., R. Debnath, J. Malik and A. Sarkar (2018). Low-income housing layouts under socio-architectural complexities: A parametric study for sustainable slum rehabilitation. *Sustainable Cities and Society*. **41**: 126–138.] for a discussion on national housing policies on low-income neighbourhoods and their implications].

5 For discussions on clientelism within informal settings, see—Mitlin, D. (2014). Politics, informality and clientelism—Exploring a pro-poor urban politics. *ESID Working Paper No. 34*: 36.

6 For conceptualisation and discussion on how western biases have dominated discourses on development and modernity, see Escobar, A. (1995). *1995: Encountering Development: The Making and Unmaking of the Third World*. Princeton, NJ, Princeton University Press and Escobar, A. (2008). *Territories of Difference: Place, Movements, Life, Redes*. Durham, Duke University Press: 456.

7 FAR (Floor Area Ratio) is calculated by the total built-up area divided by the plot area and expressed as a ratio. According to the Kolkata Municipal Corporation building regulations, in case of old buildings in a state of disrepair, the allowable FAR can be maximised by up to 50% of the original permitted FAR, and all tenants of these buildings must be rehoused in the new building. See—KMC (2017). *Assessment of Premises*. Retrieved March 28, 2023, from www.kmcgov.in/KMCPortal/jsp/KMCAssessmentHome.jsp.

References

AlSayyad, N. and A. Roy (2003). *Urban Informality: Transnational Perspectives from the Middle East, Latin America, and South Asia*. Lanham, MD, Lexington Books.

Banks, N., D. Mitlin and M. Lombard (2020). Urban informality as a site of critical analysis. *The Journal of Development Studies*. **56**: 223–238.

Bardhan, R., R. Debnath, J. Malik and A. Sarkar (2018). Low-income housing layouts under socio-architectural complexities: A parametric study for sustainable slum rehabilitation. *Sustainable Cities and Society*. **41**: 126–138.

Bhan, G. (2017). From the *basti* to the 'house': Socio-spatial readings of housing policy in India. *Current Sociology*. **65**: 587–602.

British Council (2019). Mapping the creative economy around Durga Puja 2019. British Council.

Chattaraj, S. (2016). Organizing the unorganized: Union membership and earnings in India's informal economy. *BSG Working Paper*.

Das, M. (2009). No solid roof over head, Kumartuli artisans bear with losses. *The Indian Express*. Kolkata.

Datta Roy, R. (2008). No land for idol makers; Bengal finds it hard to move workshops. *Livemint.com*. Kolkata.

Dovey, K. (2013). Informalising architecture: The challenge of informal settlements. *Architectural Design*. **83**: 82–89.

Escobar, A. (1995). *1995: Encountering Development: The Making and Unmaking of the Third World*. Princeton, NJ, Princeton University Press.

Escobar, A. (2008). *Territories of Difference: Place, Movements, Life, Redes*. Durham, Duke University Press: 456.

Hosagrahar, J. (2005). *Indigenous Modernities: Negotiating Architecture and Urbanism*. Oxford, Routledge.

KMC (2017). *Assessment of Premises*. Retrieved March 28, 2023, from www.kmcgov.in/KMCPortal/jsp/KMCAssessmentHome.jsp.

KMDA (2009). Kumartuli Urban Renewal Project.

Lombard, M. (2014). Constructing ordinary places: Place-making in urban informal settlements in Mexico. *Progress in Planning*, Elsevier. **94**: 1–53.

McFarlane, C. (2012). Rethinking informality: Politics, crisis, and the city. *Planning Theory and Practice*. **13**: 89–108.

Mitlin, D. (2014). Politics, informality and clientelism—Exploring a pro-poor urban politics. *ESID Working Paper No. 34*: 36.

Rao, U. (2018). Incremental gentrification: Upgrading and the predicaments of making (Indian) cities slum-free. *The Routledge Handbook of Anthropology and the City*. New York, Routledge: 214–227.

Roy, A. (2003). *City Requiem, Calcutta: Gender and the Politics of Poverty*. Minneapolis, MN and London, University of Minnesota Press. **10**.

Roy, A. (2005). Urban informality: Toward an epistemology of planning. *Urban Informality: Toward an Epistemology of Planning*. **71**: 147–158.

Roy, A. (2009). Why India cannot plan its cities: Informality, insurgence and the idiom of urbanization. *Planning Theory*. **8**: 76–87.

Roy, A. (2016a). Urban informality: The production of space and practice of planning. *Readings in Planning Theory (Fourth Edition)*. R. Crane and R. Weber. Oxford, Oxford Handbook of Urban Planning.

Roy, A. (2016b). Who's afraid of postcolonial theory? *International Journal of Urban and Regional Research*. **40**: 200–209.

Simone, A. (2004). People as infrastructure: Intersecting fragments in Johannesburg. *Public Culture*. **16**(3): 407–429.

Unni, J. and U. Rani (2003). Social protection for informal workers in India: Insecurities, instruments and institutional mechanisms. *Development and Change*. **34**: 127–161.

6 The emerging and diverging spaces of production

Kumartuli on a regular day

On a summer afternoon in 2018, the scene in Kumartuli was quite dissimilar to that during the peak distribution season scene presented in Chapter 4. As I walked past the workshops, I was welcomed by the damp smell of the stored clay: much more intense than the petrichor of wet earth after a torrential shower. The smell of the untreated straw and the constant hammering noise from the people at work were evocative of my earlier visits to the bustling neighbourhood. The smell and sounds were signs of the beginning of the work season over the coming busy months. This resonating sound from the workshops was sometimes disrupted by the unique call from the occasional *pheriwala*,[1] perhaps selling an item of everyday use or even an item for the decorative purpose of the idol. As I walked past the workshop-residences and reached the street crossings, the unwelcome sight and smell of public urinals and the hand-drawn waste disposal carts, not as overcrowded as during the peak season, drew my attention. I turned the street corner and immediately moved to a workshop threshold to make way for a speeding motorbike. Slowly, it became clear to me that there was a different pace to everyday life in Kumartuli beyond the peak production season. What seemed like a bustling, overcrowded place in October 2017 felt quieter, emptier, and, dare I say, cleaner in April 2018.

I exchanged a few words with a resident whom I interviewed earlier during my fieldwork. He informed me that business was slow this time of the year; he was still repairing the mezzanine floor of his workshop. Across the street, I saw a couple of tourists being guided through the neighbourhood with enthusiastic words from their local tour guide. My interviewee explained that tourists find the idol-making process interesting and often ask to witness the idol-makers at work. They click pictures and are mostly impressed by the beautifully handcrafted idols. Strangely, I was visiting today to ask for consent to carry out my day-long spatial study at his workshop. I assured him that I would not be invasive during my study.

***

Kumartuli neighbourhood has been at the centre of the idol-making industry in Kolkata for decades, and it is situated within a larger network of allied practices that form part of the grand Durga Puja business. Kumartuli may have started as the

DOI: 10.4324/9781003341222-6

idol-making hub of Kolkata because of patronage from the local *babu* families, but the locational advantage of the neighbourhood also played an important role in establishing the practices of, and allied to, idol-making here. Over time, layers of traditional ritualistic and modern arts have entangled with idol-making practice within the contested spaces of this congested *basti* situated at a well-connected riverfront location.

Heierstad (2017, pp. 25) writes,

> [I]n Kolkata's Kumartuli the stories of indigenous modernity seem to be concealed behind the tradition. Among the potters inhabiting the shanty-like, earth-floored workshops of the caste-based neighbourhood, the history of a modern and economically neoliberal-minded India unfolds.
>
> (Heierstad 2017)

The potters' humble residence (*kumar-bari*) used to be the centre of idol-making practices. The potters balanced their everyday family lives as well as constructed pots and idols within the constrained spaces of the idol-makers' quarters in Kolkata. Over the years, idol-making practices have changed. Also, the spaces used for performing these practices have adapted, grown, and evolved. The growing consumer-driven festivities and state-led promotional events are slowly modifying the related practices of assembling, producing, and distributing idols. A closer examination into individual units of idol production, storage, and residences of contemporary *kumars* implies that there has been a steady shift in the spatial configurations. Drawing on the earlier description of workshop-residences and further descriptions in this chapter, I argue that the physical spaces of production are shaped around the practices that are performed within them and continue to evolve with them.

Changing spaces: repurposed workshop

A new type of workshop is emerging in the existing buildings on the *Majhergoli* and narrow parallel lanes in the core area of Kumartuli. These buildings are being modified and transformed internally to serve as workshops instead of their original purpose as workshop-residences (Figure 6.1). The owners or the main artists, who lived in these houses while also using the front rooms like workshops, are moving their families to nearby places and using these buildings as workshop-cum-storage-cum-accommodation for the seasonal migrant workers. Although the size of these buildings remains like the original workshop-residences, these buildings now serve as standalone workshops.

This particular workshop was studied at a different time of the year. I have reflected on this study to illustrate the seasonal working pattern. During the mapping exercise in April 2018, this workshop was almost empty, and one of the two men mentioned in the study was preparing the frames for the start of work. However, this workshop was well-stocked and brimming with preparatory activities during the autumn, just after the Durga Puja.

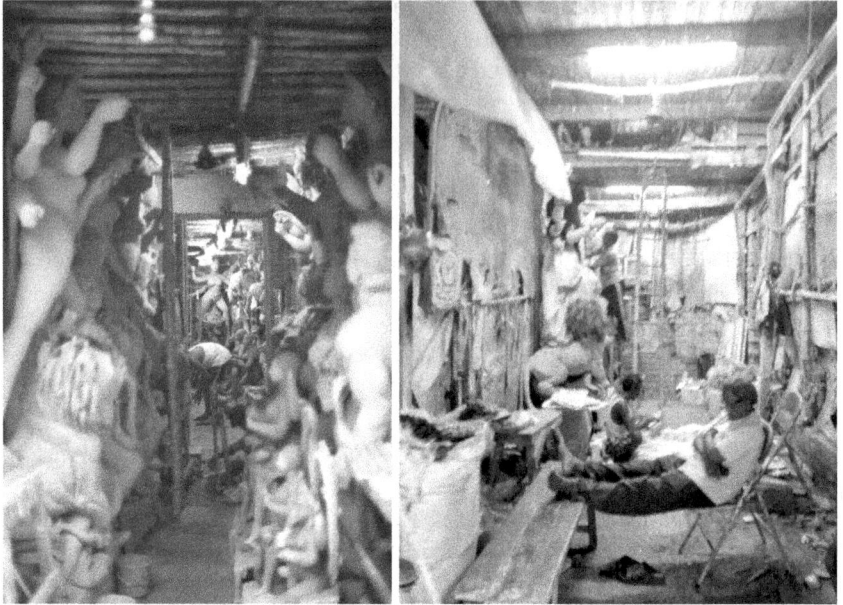

Figure 6.1 Repurposed workshop with front-room workshop and storage spaces

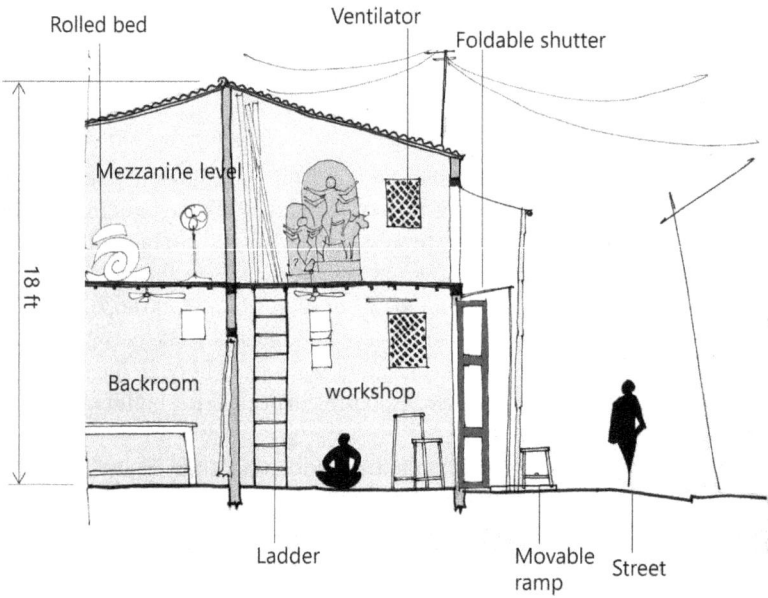

Figure 6.2 A section of a repurposed workshop

A few days before *Lakshmi* Puja 2017 (October), I interviewed three artists (interview: Kumartuli 2017) in their workshop, which currently is a standalone type (Figure 6.2). One of my first interviewees was a woman entrepreneur. I sat opposite her on a footstool and began consciously recording her. She carried on working on the smaller Lakshmi idol, glueing the hair and occasionally picking the idol up to check the symmetry, balance, and overall appearance—the all-important appearance and aesthetics that would ultimately sell the idol. I recorded all three interviewees' movements and activities within a time frame of three hours.

The workshop's basic plan (Figure 6.3) is similar to workshop-residences (discussed in Chapter 3) with a front room, half partition, a back room, and a *macha*. A toilet, also shared by other families, is accessed through the back door of the house. The three artists—the woman, her husband, and her brother—have been continuing their father's business from this house. A few years earlier, the couple had moved out when they could finally afford to buy an apartment for themselves

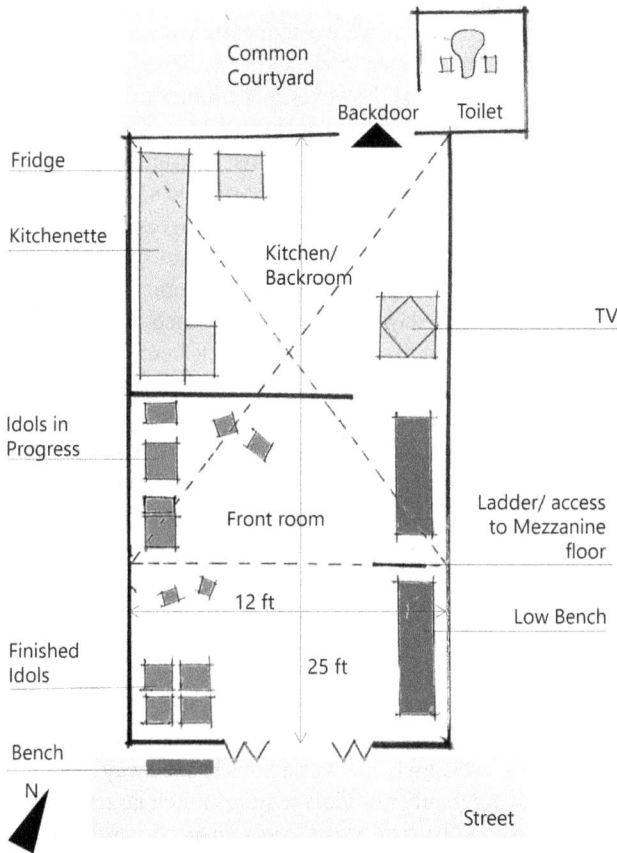

Figure 6.3 Details of a typical repurposed workshop

just outside Kumartuli. However, her brother still continues to live in a rented room just behind this house. After the busy Durga Puja season, they were selling idols for the Lakshmi puja while also working on them at their workshop. All three of them sat close to the sidewall with their backs to the wall and had rows of small idols, mostly unfinished, in front of them. On the opposite sidewall, a couple of small footstools were kept where customers could come and sit while their idols were being finished. I occupied the innermost position in the workshop while I interviewed and watched a couple of customers come in to buy idols. Some interested parties saw from outside, while a few entered the workshop to have a closer look.

A lady came in, chose an idol, commissioned it, and indicated the artist to add the finishing touches to her taste. The woman was the main artist who negotiated prices and convinced the customers to buy from her. The buyer insisted on bargaining the price; however, the 'small yet pleasingly decorated idol' was not for sale at a lower price. Fifteen minutes later, she walked out of the workshop with the idol and a satisfied look on her face. I later discovered that the female artist is influential in the idol-makers' representative group and their cooperatives. Her position within the workshop was not gendered at that moment while she was selling *Lakshmi* idols for domestic festivities. Customers were also mostly women. She remained the key artist and salesperson, perhaps to tap on the women clientele for the domestic ritual of *Lakshmi* puja. For Durga Puja, however, her brother and her husband take the lead in crafting and sales. This particular business has earned recent fame due to the lady's appearance on television as a female-artist in Kumartuli. However, in general, women's role remains in the household chores and serving the male members of the family. Her sister-in-law (brother's wife), who lives in a rental accommodation just behind the workshop, occasionally came into the workshop during my study to help by passing things and cleaning. At other times, she remained involved in household chores and everyday rituals. Close to lunch, the female artist quickly packed up, went to the back office, and rushed back home nearby to cook for the day's meals, meals that her family would be having throughout the day.

After she left, the two men continued working on and selling the idols. The two men took a few breaks in between and resumed work. As I left the room, I saw the other lady rearranging the footstools and cleaning the space. This activity, I presumed, was a recurring one, a result of customers walking in with their muddy shoes. Ideally, artists would always want their shopfronts and workspaces to be in order and clean to appeal to potential customers.

The size and height of the repurposed workshops are not as big as the 'factory-shed' workshops. Therefore, artists have to rent storage spaces outside Kumartuli to accommodate larger idols after construction. Idols are moved outside the workshop to the storage spaces nearby and brought back for selling. This process reflects the changing needs and demands of the consumers. As the space requirement for the idol-making industry grew, slowly the workshops broke away from the residences. Additionally, the growing number of idols requires much larger production or storage spaces. This also implies that artists can rent need-based spaces supporting their thriving businesses. Artists balance their production and sales spaces over residential and storage spaces to remain closer to the branded hub of Kumartuli.

Holding on to the smaller workshop spaces to do business from Kumartuli proves the implications of the place-based connections, ease of business, and the significance of the Kumartuli-brand name. This new type of workshop shows an emerging pattern of practices deeply embedded in place and rooted in more productive functions of idol-crafting.

Agency and new typologies

Before being stalled, the KMDA redevelopment project that was proposed for Kumartuli had already started to relocate some families to a nearby warehouse at Kumartuli 2 in 2011. This relocation would have been temporary until their residence-workshop was rebuilt. Demolition work for this had started in 2011, and some workshops on one side of Kumartuli (near Kumartuli Street) were affected. However, due to collective resistance to the proposed redevelopment project, work was stalled abruptly, and, eventually, the plan was scrapped with no further action taken by the government on these affected buildings or families. As a result, while most of these relocated families continue to work from the nearby warehouse, they still have retained their original spaces in Kumartuli. When KMDA left the project, some buildings had already been demolished. The *maliks* of these buildings since 2011 used temporary sheds to continue their practice, mainly the client meetings and initial transactions during peak seasons. These temporary sheds were only offices, while idol-making activities went on in the more permanent structure of the warehouse.

During the summer of 2018, through my daily visits, I was able to witness the permanent rebuilding of one such workshop. The main artist paid for the project. Interestingly, even in 2018, they chose to build the workshop with the simple traditional layout—a small room at the back and a bigger room in the front. The walls were 20 feet high with a timbre and plywood *macha* at the 12-foot level along the entire length of the building. A bamboo ladder would provide access to the macha, and the roof was being made of corrugated tin sheets (Figure 6.4). The seasonality dominates the idol-crafting and delivery schedules; hence even while the construction of the workshop was underway, partly sculpted idols were also within the workshop space, suggesting work-in-progress. By the end of April, the building was complete and ready to commence work for the upcoming season. Therefore, the owners of this workshop, the father and son, who are now both involved in the family business, would have two workshop-cum-storage spaces. The family would also continue to live on the repurposed first floor of the warehouse.

Appropriation and socio-spatial relations

The spatial study reveals important features of how workshop spaces are used and inhabited. The front facades of all the buildings are similar, irrespective of being old or new, containing foldable wooden framed tin/aluminium doors; plastic sheets serve as awnings supported by bamboo during the monsoon months. Precariously built benches are work surfaces, and all the main artists follow traditional everyday

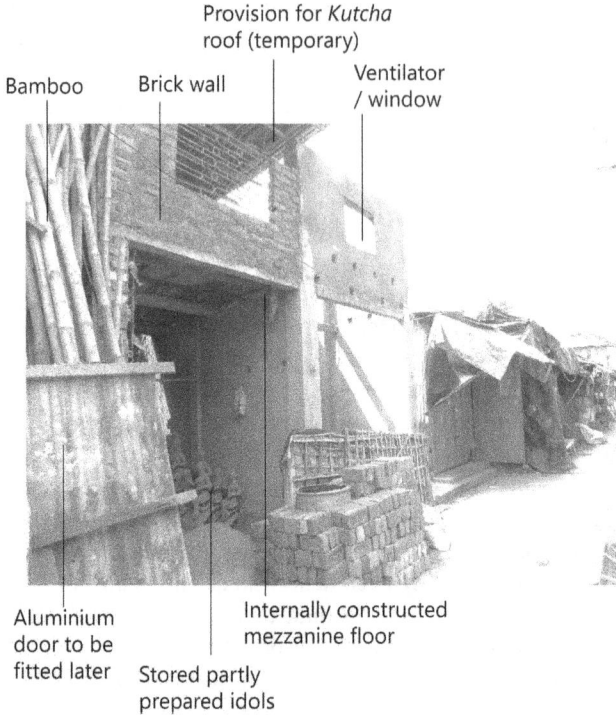

Provision for *Kutcha*
roof (temporary)

Bamboo   Brick wall   Ventilator
/ window

Aluminium
door to be
fitted later   Internally constructed
mezzanine floor

Stored partly
prepared idols

Figure 6.4 New construction of workshop (April 2018) on the plot of demolished workshop-
        residence by KMDA

rituals related to their idol-making practices. Nevertheless, by no means are all
the spaces of production the same; they are rather heterogeneously built. Most are
shaped by practices related to the individual idol-maker's enterprise, flows of materials and labour, and the elements like competencies that shape these practices. The
repurposed and new-rebuilt workshops illustrate how the Kumartuli neighbourhood is slowly transforming into commercialised idol-producing units from age-
old workshop-residences. However, this study also suggests that build spaces are
unlikely to change significantly unless triggered by an external cause. To elaborate
on this argument, in the following paragraphs, I present a comparative spatial analysis of the three types of workshop spaces discussed in this book (Chapters 3 and 6).

The spatial study of the three types of workshops suggests particular similarities
in the use of the workshop space: all the primary idol-crafting work are conducted
on the floor. In the (first) workshop-residences, the seasonal workers were binding
the straw to sculpt the limbs of the idols. They use the floor as the plane to position
the straw in place while using their feet and arms to tighten the straw and tie them
into a cylindrical shape. The (second) factory-shed workshop was much larger and
accommodated more workers and idols at a time. Here, the carpenter also place
the timber and tools on the floor and cuts the batons to size with his saw while

holding them into position using his feet. In the (third) repurposed workshop, all three artists were working while sitting on the floor and finishing the smaller idols. All these workshops use the mezzanine floors primarily to store the smaller idols. Additionally, these spaces are large enough to house seasonal workers during the idol production phase.

From the comparative spatial study and fieldnotes, it is evident that there is a hierarchy of users in the workshop spaces. In the case of the workshop-residences (first), the woman artist's movement suggests that she uses the interior residential space as well as the workshop space. However, the movements of the seasonal workers are only restricted to the workshop and mezzanine floor. They use toilet facilities outside the house. In the case of the 'factory-shed' (second) workshop, even though the unit is not attached to a residence, there is still a hierarchy of users within the workshop spaces. The office space is restricted to the main artist only, which remains locked when not in use. The cooking area at the back of the workshop is used mainly by the trainee seasonal worker. During my study, the main artist was working on a tabletop and sculpting finer details of the finger and toes of the idol with soft clay, while the trainee worker was responsible for the labour-intensive work or running chores. I did not notice much hierarchy in the (third) repurposed workshop type, as all three artists were relatives and enjoyed the some-what same social status. The analysis illustrates a long-lasting trace of the spatial structure of the place and shows how spaces, while remaining somewhat flexible, are organised according to codes of practice by users. Despite the different typologies of workshops presented in this book, I must stress that the multiplicity and heterogeneity within the built spaces of Kumartuli are immense. Typologies are only based on the primary use of each such space and the time when these workshops were built. Other than that, in all three types of workshops, there remain underlying interconnected elements of the practice of idol-making through everyday ritualistic norms that have been carried on through generations, such as cleaning the space routinely in the mornings and evenings and lighting incense sticks. All four types of workshops, big and small, conventional and converted, still have designated spaces for crafting idols and storage, cooking and eating within limited resources, and housing the seasonal workforce.

The study of the building types in Kumartuli raises questions not only of the capacity of the idol-maker to build their workshops but also of certain ownership dualities that restrict the construction of new buildings. Also, the question remains whether the small, fragmented plots of land resulting from years of subdivision would have the potential of a new building development with modern building regulations.[2] The new building regulations require following certain norms and setbacks guided by the plot size and location. These restrictions perhaps would negate any usable building. Also, the inner-city location raises questions of real estate prices, speculations, and propriety. It is therefore related to the ownership issues, the tenancy types, and certain slum legislations (discussed in Chapter 5) in order to analyse the built nature and socio-spatial transformation of the neighbourhood.

Spatial factors cannot be considered in isolation from the economy, class, and gender. Massey argues that '*broader social structures of community, changing*

*patterns of consumption . . . changing national ideological and political climate
and the marked patterns of geographical, cultural differentiation—all of which'*
contribute to the uneven spatial development when she compares industrial and
social landscapes of changes (Massey 1995, pp. 188). Empirical results show the
spatial and economic reconfigurations within the neighbourhood, driven by the
changing pattern of consumerism, resulting from rising global demands and media
showbiz towards this traditional craft. Reflecting on Massey (Massey 2005), the
spaces within the workshops should not be considered for analysis in isolation
but rather as part of a larger network of flows that affect production processes.
The study of the different types of workshops reveals a process of emergence, the
making and remaking of spatial configurations that underline the ongoing transformation of Kumartuli. The study highlights the socio-spatial characteristics of
Kumartuli's buildings and, more importantly, the unique production spaces. The
availability of labour, materials, and orders for finished products largely governs
the geography of the place. Hence, I argue that individual units of production are
spaces shaped by the practices performed in them as well as by the multiple relations of the users, labours, material, and finished goods. In other words, not only
are the individual spaces of production largely shaped by the modes of investment
and social positions manifested through infrastructural configurations, class, caste,
and gender identities, as noted in the literature on relational geographies (Truelove
and Cornea 2020), but also by the elements of practices such as materiality, com-
petences, and technical know-how.

The spatial study and analysis of user movements suggest that the workshop
spaces are used in conjunction with the street in front, which is part of the personal realm of the workshop-owner as well as a vibrant public space. In addition to
the productive functions of the workshop, the street is a seasonally transformative
space that fulfils productive and commercial functions as well as a space for social
and cultural exchanges at other times. The spatial dominance behaviour over the
street is observed in the owner/primary occupier of the workshop-residence more
prominently than the hired seasonal worker, suggesting a higher social position of
the *malik* (owner of the business). The floor spaces are more important than the
walls, while also the mezzanine floors, built from more temporary materials, are
used for storage and sleeping. Even if the spaces are cramped and contested, the
practices are more or less grounded on the floor to provide stability and ease of
business and unobstructed access to the street.

The movement diagrams reveal the spatial practices and how they relate to
the social and cultural aspects: for example when the workshops were part of the
household, roles were gendered, and the woman continued to work at the back. In
the changed scenario, the woman still goes back home to cook for the family and
returns to work only in the afternoon. In contrast, the woman in the conventional
type of workshop-residence can take the lead due to her attached residence; she
has been managing both the household and the business flexibly due to the built
character of her production space. So, women's social position is not particularly
gendered but varies with circumstances. Some women have also earned acclaim on
the basis of their intricately sculpted idols. However, her social position dictates
her movement within the production unit and her command over the space. This

perspective is reflected even within the idol-making industry's political economy. With the changing pattern of consumption and production practices, new norms dictate the hierarchy of the idol-making practice.

While a sizeable proportion of the total area of the front rooms is actually taken up by the users, both the employer and the employees, the backrooms remain within the domain of the *malik* and their family. The 'factory-shed' standalone workshops are also spatially dominated by the workshop's owner, while the 'repurposed' workshops showed an uneven dominance over the space even within the same family of *maliks*. However, I want to highlight that individual workshops are unique spaces of practice and products of differential economic, social, cultural, and political arrangements. Even using a Southern lens, such spatial variegations are difficult to fully conceptualise considering the multiplicity and heterogeneity of the practices and produce a uniform template of the production spaces. Hence, a cookie-cutter template by the redevelopment plan might not have served the residents.

The aftermath of the redevelopment plan and the government's policies are slowly changing the nature of the neighbourhood: what started off as a caste-based homogeneous neighbourhood is now a commercial cluster of idol-making workshops, most of these differently materialised to suit individual requirements. Through the seasonal trough, some workshops and streets shift considerably spatially: the streets are used for communication, while the workshops are overlaps of related idol-making practices and the everyday realities of family living. During the seasonal peak, however, some residences and workshops, in addition to housing families and *kumars*, also house seasonal migrant workers. The streets are lined with overhangs and extensions of workshops and are spaces of sociocultural exchanges and increased economic activities triggered by the consistent seasonal flow of migrant workers, tourists and clients, raw materials, and finished products. There is striking seasonality in the character of the place, which changes phenomenally during busier and slower phases.

Spatial flexibility and reparation in a Kolkata *basti*

Practices are shaped largely by the socio-economic and political constructs of the locations that they are performed in (Khalid and Sunikka-Blank 2017). An interesting take on the practice theoretical perspective is to draw on the relations between spatiality and practice. Analysis of workshop requirements through interviews and photographs on the connection to the spaces of production is in line with the materiality of the space, the size, the height, and the built character. I draw upon the innovative and reversible modes of construction of the workshop and residential units. The mezzanine floors are frugally built from wood and bamboo, which largely differ from Kolkata's standard brick and concrete construction. Further, these spaces are flexible and provide scope for easy extensions. The doors and windows of the front façade of workshops provide flexibility. It is also imperative from the study of workshops and relevant practices in Kumartuli that ground floor locations provide easy access to the streets. However, the spatial study opened up questions of the dynamics of the spatial configurations governed by social and economic positions. Therefore, it might be fair to say that practices in Kumartuli are influenced by age-old socially constructed norms and meanings of traditional idol-making practices coupled with the changing materiality and the technical competence of the practitioners due to increasing investments and consumer-driven commodification of the idols.

Reflecting on the earlier discussion presented in relation to the KMDA plan, I revealed that certain younger and more affluent groups of artists presented views on modernity quite in contrast to the existing auto-constructed workshop spaces. They suggested better, brick and mortar built, higher workshops to accommodate the changing needs and requirements. These views indicate that due to the changing demand and production functions, necessary spatial changes might be required to accommodate these. For example in the case of repurposed (third type) workshops, the *kumars* have modified and repurposed the spaces to accommodate the growing needs. However, the spatial study consolidates and implies the broader socio-political relations that impact the creative and productive functions and the resultant spatial transformations.

The socio-spatial character of the *basti* neighbourhood often mars the unique creative and economic identity of Kumartuli as a place. However, I argue that cultural and material practices and the multiple social and economic relations shape the spatial character of a place. A nuanced understanding of spatial and material practices and flows is important to produce unbiased views of such marginalised neighbourhoods. Further, Banks et al. suggest that studies on urban informality remain related and bounded in academic disciplines (Banks et al. 2020). A more in-depth analysis of the multiple facets of the production and creative functions of the neighbourhood opens discussion on the political economy of this informal sector industry with relation to the changing socio-spatial characters. This relates to furthering the understanding of land, buildings, and spaces within informality, which are in the process of change: sites where relations are formed.

Urban planners and geographers are increasingly keen on discarding the idea of a uniform or singular urban form and acknowledging diverse socio-economic, socio-spatial, and multi-scalar territorial formations through the process of urbanisation, quite contrary to the previous normative forms of spatial development driven by the idea of a bounded city (Brenner and Schmid 2015). Developing on the idea that places have to be thought in the relational perspective (Graham and Healey 1999)—we cannot consider the place value of Kumartuli in terms of only the real estate value of Kumartuli's central inner-city location or connectivity or whatever the residents perceive but in the wider contexts of the global geographies the idols produced in Kumartuli shape.

Kumartuli as a place might have once been a singular neighbourhood at the centre of the idol-making industry, but it is no longer a singular entity. Kumartuli developed as a singular site as a brand for idol production in Kolkata; however, I argue that a number of wider relations facilitate the development of a network across multiple geographies that support the industry. The community representatives and stakeholders echoed this view during the deliberative workshop in 2022. Practices in Kumartuli shape wider socio-spatial geographies through local and global flows. Also, seasonality and social, cultural, and political relations shape the spaces of Kumartuli. Spaces are under transition and continue to emerge with multiple meanings and a growing network of relations. Residential spaces are continuously transformed into productive spaces while multiple trajectories of change continue to shape the places. I reflect on Roy (Roy 2009) and McFarlane (McFarlane 2012) in presenting the flexibility in informal spaces under the transformation process. These spaces need to be understood with reflections on the wider implications of multiple actors navigating these spaces through market-driven appropriation and bringing about the uneven suspension of legalities, hence deregulated growth (Banks et al. 2020).

Through a Southern lens, cities should not be considered as the 'privileged' concrete sites of familiar local or metropolitan units. The broader, more heterogeneous landscape of urbanisation is better associated with flows and networks across a wider socio-spatial scale (Brenner and Schmid 2015; Lawhon et al. 2018). The cartographies that define Kumartuli's idol-making industry are characterised by multiple geographies of migration, flows of materials, and finished products both locally and globally. While the uniqueness of Kumartuli is in the place-based character, it is often marginalised due to the appearance of a slum and the relation of caste and class. Postcolonial urban theorists too often problematise informality in a generalised manner (Roy 2016). The empirical data, however, illustrates that in the case of Kumartuli, even within a perceived slum neighbourhood, there are multiplicities of practices that shape spaces of living and working with seasonality. The study into the spatial practices of Kumartuli has opened up relational understandings of flow and materialities that affect practices.

Southern urban practices and reference to squatting, consolidating, and building upon as practices are processes of this significant phenomenon of urban informality (Bhan 2019) and are dependent on land deeds and wider relations of businesses. Despite the land and building ownership complexities and a subsequently failed redevelopment plan due to this, spaces in Kumartuli are in high demand. More established idol-makers rent more workshop spaces, and families of other residents or struggling idol-makers are forced to move elsewhere, producing unequal modes of reconstruction. This results in rising disparity and social inequality due to the extensive commercialisation of the neighbourhood. Although culture-led initiatives have been linked to economic redevelopment (Kanai and Ortega-Alcazar 2009), there are more complexities related to cultural-industries-led place-based regeneration.

Displacement and gentrification are also due to the commercialisation of the neighbourhood. Ghertner questions the viability of generalising gentrification and argues whether areas with predominantly slum-type land tenure (in the case of Indian cities) could be accounted for as cases of gentrification (Ghertner 2014). In the case of Kumartuli's displacement issue, whether or not the land/building is used for better economic value is a question. That could be answered from a relational perspective. In summary, I urge for more examination of the multiplicity and the temporality of the spaces of production in the idol-making practice in order to understand the adaptability of practices with continued spatial constraints.

Notes

1 A mobile salesperson calling out for his sale carrying in a basket on his head or a cycle van.
2 Contemporary building regulations in India follow strict guidance on covered areas, setbacks, and fire safety codes. KMC (2017). *Assessment of Premises*. Retrieved March 28, 2023, from www.kmcgov.in/KMCPortal/jsp/KMCAssessmentHome.jsp.

References

Banks, N., D. Mitlin and M. Lombard (2020). Urban informality as a site of critical analysis. *The Journal of Development Studies*. **56**: 223–238.

Bhan, G. (2019). Notes on a Southern urban practice. *Environment and Urbanisation*. **31**: 639–654.

Brenner, N. and C. Schmid (2015). Towards a new epistemology of the urban? *City*. **19**: 151–182.

Ghertner, D.A. (2014). India's urban revolution: Geographies of displacement beyond gentrification. *Environment and Planning A: Economy and Space*. **46**: 1554–1571.

Graham, S. and P. Healey (1999). Relational concepts of space and place: Issues for planning theory and practice. *European Planning Studies*. **7**: 63–646.

Heierstad, G. (2017). *Caste, Entrepreneurship and the Illusions of Tradition: Branding the Potters of Kolkata*. London, Anthem Press.

Kanai, M. and I. Ortega-Alcazar (2009). The prospects for progressive culture-led urban regeneration in Latin America: Cases from Mexico City and Buenos Aires. *International Journal of Urban and Regional Research*. **33**: 483–501.

Khalid, R. and M. Sunikka-Blank (2017). Homely social practices, uncanny electricity demands: Class, culture and material dynamics in Pakistan. *Energy Research and Social Science*, Elsevier. **34**: 122–131.

KMC (2017). *Assessment of Premises*. Retrieved March 28, 2023, from www.kmcgov.in/KMCPortal/jsp/KMCAssessmentHome.jsp.

Lawhon, M., D. Nilsson, J. Silver, H. Ernstson and S. Lwasa (2018). Thinking through heterogeneous infrastructure configurations. *Urban Studies*. **55**(4): 720–732.

Massey, D. (1995). *Massey D. Spatial Divisions of Labour: Social Structures and the Geography of Production*. London, Macmillan International Higher Education.

Massey, D.B. (2005). *For Space*. London, SAGE Publications: 222.

McFarlane, C. (2012). Rethinking informality: Politics, crisis, and the city. *Planning Theory and Practice*. **13**: 89–108.

Roy, A. (2009). Why India cannot plan its cities: Informality, insurgence and the idiom of urbanisation. *Planning Theory*. **8**: 76–87.

Roy, A. (2016). Who's afraid of postcolonial theory? *International Journal of Urban and Regional Research*. **40**: 200–209.

Truelove, Y. and N. Cornea (2020). Rethinking urban environmental and infrastructural governance in the everyday: Perspectives from and of the global South. *Environment and Planning C: Politics and Space*. **39**(2): 231–246.

7 Kumartuli's future?

Kumartuli's present

The festival of Durga Puja in contemporary times is not confined to only idol-making, although the idol remains central due to the religious and material cultural heritage associated with the festival. Despite the associated practices and elaborate artistic imaginations of the idols and the *pandal*, the meaning of the festival to the general public is still linked with the idol-worship, central to any Hindu ritualistic practice. Moreover, in Kolkata, most idols annually prepared for famous contemporary celebrations have humble beginnings from the ancestors of the present Kumartuli potters to this date and carry the brand of the respective family and Kumartuli. The occupation, which was only confined to the potters of Kumartuli and their families, has gradually spread among other urban artists who are building their own brands loosely associated with Kumartuli.[1] Kumartuli is in the centre of the idol-making industry drawing in raw materials and the seasonal migrant workforce. Over time, although other smaller idol-crafting settlements in Kolkata and adjoining areas have grown, Kumartuli remains one of the most prominent and the largest in terms of the number of businesses and area. Moreover, the centrality of Kumartuli in idol-making and associated crafting continues to be unchallenged despite redevelopment pressures, infrastructural precarities, and cramped working conditions. Hence, the locational advantage, relational geographies, and branding of the place itself have fostered the network within it and strived to maintain its identity through the history of Kolkata.

A new group of trained artists have started sculpting modern renditions of the Durga idol, differing largely from the traditional conventions. The emerging group of artists receive many of the contracts for the bigger-budget Durga Pujas of Kolkata due to the pressures of earning accolades. Hence, promoting Durga Puja as a tourist-driven entity threatens the traditional livelihoods of the caste-based artists of Kumartuli. The immense political patronage risks the festivities being completely driven by market forces. I continue investigating the relational perspectives that shape cultural practices and inform socially inclusive policies.

Influenced by the postcolonial debates on modernity and culture, it might be argued that the subaltern, in this case, the unrecognised contribution of the idol-making community, is not quite considered in the mainstream class and caste

DOI: 10.4324/9781003341222-7

discourse, rather, it is constantly marginalised (Chatterjee 2011). The Kumartuli artists, belonging to particularly lower castes of the Hindu society, produce cultural materials for a thriving traditional industry at the city's heart. The notion of using culture and heritage as commodities to promote tourism and involving the regeneration of certain areas is also not uncommon. However, does the current policy address the community as a whole? Or does it support an acclaimed few who are culturally equipped to promote their businesses? While the creative economy framework might be a tool of the neoliberal economy, it is important to analyse how the idol-making industry could be affected by it. I questioned through this book whether the cultural industry framework and the westernised notions of modernity pose significant challenges and continue to marginalise the existing practices of Kumartuli.

The social, cultural, and economic practices of the Kumartuli neighbourhood are very much contradictory to modern living. However, globalisation and post-colonial thinking govern the idea of modern living (lifestyle as well as the built environment) among the neighbourhood residents. The colonial influence still governs the city's practices in establishing authority, legitimacy, and identity.[2] Similar to the informal nature of land tenure and the built environment, the idol-makers of Kumartuli are part of the informal sector economy. The practice of idol-making is informal and seasonal. However, the idol-making industry supports a wide network of the labour force and is part of a successful cultural practice that directly and indirectly contributes significantly to India's local, regional, and national economy.

The practices in the Kumartuli neighbourhood align with the traditional norms, far from the conventional perceptions of neighbourhoods of a western city. As noted in contemporary works (Heierstad 2017), Kumartuli is a lucrative area to promote the tourism industry and churn foreign revenue through tourism, but the tourism policy does not address the existing practices of the Kumartuli neighbourhood. McFarlane and Silver (McFarlane and Silver 2017) call for research on everyday life, social and cultural practices, and the development of urban infrastructure focusing on these. Based on people's everyday choices and the place-making within their communities, it might be possible to make socially and culturally inclusive urban spaces to avoid marginality through policy frameworks.

In India, the general public discourse of being modern relates to something western, much influenced by the colonial period. Caste is often contradictory and almost an antonym to modern ideas, but *kumars* use caste as a tool to promote their businesses (Heierstad 2017). In a way, the community of Kumartuli idol-makers are part of a caste-based settlement whose interwoven social and cultural practices have nurtured a highly lucrative cultural industry over a considerable period of time. In the process, the once informal settlement has been registered as a *basti*, and densely self-built houses and workshops characterise this. The current trend of spectacular consumerism and simultaneous cultural tourism policy patronage are strikingly changing the nature of idol-crafting and the relevant festive installations associated with Durga Puja (Guha-Thakurta 2015).

While presenting the background of this research in the introductory chapter, I set out some reasons for the significance of this informal sector industry to the socio-political scenario of Kolkata. While the enlistment to UNESCO's ICH list had not been realised during my fieldwork in 2017 and 2018, governmental policies involving promoting cultural industries, local festivals, and tourism were already in place (Guha-Thakurta 2015; British Council 2019).

Using postcolonial theory as a critical lens, I have studied residents' place-making activities and attachment and discussed the various places, like the riverfront *ghat* (embankments), alleys, temples and individual workshops, in the neighbourhood that residents associate meanings with and feel connected to. In addition to showing residents' place identity, this study also presented how they think the idol-making cluster in Kumartuli is embedded in place and benefits from the evolved multimodal network. In answering the central question, the book highlights the process of transformation of the place and acknowledges earlier theorisations on informality, which recognise the governmental appropriation and legitimisation of certain informal settlements (Roy 2009; McFarlane 2012). In this case, despite the informal tenure, many similar inner-city slums of Kolkata were legitimised during national slum clearance acts (Shekhar and Shekhar 2021). The interviews and the participatory visual study suggest that residents perceive the informal nature of the cluster and the frugally built form as part of the material and spatial elements of the idol-making practices. However, interviews with planners and politicians reveal an aspect of stigmatisation associated with the residents of Kumartuli.

Through the different chapters, I have presented the ongoing spatial change in the neighbourhood. While a planned intervention by KMDA and the government was resisted, there have been individual attempts to spatial reorganising and continued displacement within the neighbourhood and a significant widening of the geography of the idol production. Also, the aftermath of the plan and the government's neoliberal policies are slowly changing the homogeneous nature of the neighbourhood: what started as a caste-based homogeneous neighbourhood is now a commercial cluster of idol-making workshops, driven by faith-led consumer demands, most of these differently built to suit individual requirements. The study of the spaces of production showed four distinctively different types of workshops, and most are already transforming to accommodate the growing demands. While a few family-run traditional workshop-residences remain, many such buildings have been turned into solely idol-crafting and storage spaces. Some artists have moved their residences elsewhere. Many workshops are built primarily for idol-crafting, storage, and housing the seasonal workforce. Some workshops affected by the stalled redevelopment plan are also slowly being rebuilt. All four types of workshops continue to function as idol-making spaces, while long-term storage functions are being shifted elsewhere by artists to accommodate the growing production functions. The alleyways and street corners serve important interactive public spaces and form part of a rich social network.

After discussing the reasons for the failures of the KMDA plan and further investigating the spatial and material practices of the idol-making neighbourhood, I argue that the plan failed to address these practices and fit them with their current socio-spatial configurations into a workable building proposal. There remained a lack of communication between the users of the proposed buildings with planners and politicians. Beyond the resistance to the plan and the change in the local government, there seems to be no government intervention regarding the future of the displaced residents due to part relocation of the project. Instead, Kumartuli has been apparently left alone to recuperate from the shock of temporary relocation and an abandoned redevelopment plan. The local government has instead focused on the wider economies of the cultural productions associated with the Durga Puja celebration. In conclusion, I argue as to how the informal tenures, slum location, and disjointed infrastructure continue to marginalise the community of Kumartuli despite individual agencies in rebuilding and reshaping workshop spaces to accommodate the growing demands.

Reparations and public services

With the listing of Durga Puja as UNESCO's Intangible Cultural Heritage, it is anticipated that Kumartuli will continue to experience large-scale spatial change, gentrification, displacement, and/or overstretched municipal services and infrastructure in states of disrepair. Hence, it requires urgent policy and planning interventions from the relevant authorities to address these issues.

During the deliberative workshop, organised in 2022, in-depth discussions were held on the existing and future infrastructure and services' scenario, including possible engineering solutions for the Kumartuli neighbourhood and how participatory and inclusive planning approaches can lead to overall sustainable and equitable development of informal settlements while providing linkages between allied crafts industries in and around Kolkata's inner-city wards. Participants presented their opinions on various issues such as water supply, sanitation, drainage, road networks, and solid waste management and discussed future policy interventions and scope of further developmental works. However, participants also stressed more on the need for improving the working conditions and developing policies for better services and facilities for aiding increased production and continuing Kumartuli's heritage of idol-crafting and allied crafts and the associated brand name.

Based on the key findings of the deliberative workshop, I call for repair and maintenance. First, to better situate the position of the community in Kumartuli and reinforce the earlier investigation, I first asked whether they would consider a redevelopment project to improve the built environment and associated infrastructure and services in Kumartuli as an architect, planner, or developer's imagination would render. However, the community stakeholders in Kumartuli believed shifting, if required, should be done without disrupting the community's livelihoods. Hence, in situ improvement of roads, buildings, and existing infrastructure is the only way they could foresee. They highlighted the extreme reliance on the existing material networks and physical networks as well as allied crafts clusters and reiterated that any relocation would disrupt or displace the existing supply chains. They repeatedly stressed that the improvement of working conditions, including provision for housing seasonal labour force, was of utmost importance. Also, this should be done without hampering the seasonal work cycle of the crafting practices and communities while ensuring the continuity of Kumartuli's heritage of idol-crafting and allied crafts.

My argument stems from this key finding that the community in Kumartuli should be considered in relation to the allied crafts and networks associated with Durga Puja that has evolved over centuries. Here, as well as the existing physical and material networks, people associated with the crafts are also important infrastructure components that contribute to provisioning services for the festival economy. Considering this, I first define what repair and maintenance presented in the literature on Southern cities mean (Simone 2004; Bhan 2019). Repair refers to incremental and everyday forms through which cities and landscapes are held together through processes of fixing, mending, and maintaining (Millington 2019). The key to repair is that it refers to the built environment and infrastructure that already exist and that need continued improvement through ongoing, often overlooked work of those who maintain, modify, and repurpose the critical systems that strengthen urban lives and livelihoods. Practices of repair may provide possible pathways for ethics and politics going forward. Repair takes uncertainty and breakdown as starting points and develops forms of action that are predicated not on that which could exist but rather—on that which already exists. However, mostly repair and maintenance literature highlights cases of huge engineering works and structures where larger groups of people and labour are associated with the everyday reparative and improvement works. In the case of the Kumartuli neighbourhood, I have presented the possibilities and promises of an informal neighbourhood situated at the centre of the idol-crafting and allied industrial network and how despite the degrading conditions of physical infrastructure, it has emerged as the hub that connects through people, materials, and networks.

Through the various issues that require maintenance and improvement included mostly municipal services. For example particularly during the seasonal surges when about 1,200 to 1,500 additional people need to be accommodated in addition to the existing residents, facilities such as public toilets, waste disposal bins, waste clearing services, and water supply, including drinking water, need to be much more than normal times. Not only is it due to the surge in temporary residents, but also this is since seasonal migrant workers, employed on a temporary basis, gather in a notified slum. Already there are not enough and adequate facilities for toilets in the neighbourhood, but services are stretched due to the population surge. Through the discussion held during the workshop among the planners, engineers, and architects, stakeholders agreed that specialised interventions are required for neighbourhoods such as Kumartuli, where additional productive and commercial functions stretch municipal service limits. Services such as solid waste management and the provision of public facilities like toilets and drinking water should be integrated within localised place-based decision-making processes. Particular attention to informal neighbourhoods such as Kumartuli is urgently required to ensure the continuation of the heritage.

Contributions and implications of this research

Reflecting on the empirical findings, this research supports that, first, practices are understood in relation to the changing spaces and the place. Equally, practices shape spaces, even the constrained spaces in the densely built inner-city areas of the global South. This book traces the relationship between material relations, everyday social practices and spatiality, and the role of place in fostering them. This is observed through ethnographic research of the everyday social practices and material processes within the spaces of production of clay idols in a *basti* neighbourhood of Kolkata. The theoretical contribution of this research is in advancing our understanding of how relational spaces are shaped by networks of associated productive functions in densely inhabited areas and how interwoven social and material practices can both contest and help remake these spaces. Further, this book assesses relations in an informal setting, which reconfigure spaces continuously to accommodate the changing needs of existing practices due to the implications of public policies and consumer-driven commodification. This research discusses how an informal sector industry has been thriving despite territorial constraints, remaking and reassembling its local and international geographies to adapt to growing demand.

Through the empirical chapters, this book reflects on the place-based network of actors; spatial practices; and social, economic, and political scenario of an inner-city neighbourhood largely dependent on their creative capital. Empirical results show the spatial and economic reconfigurations within the neighbourhood, driven by the changing pattern of consumerism and rising global demands towards this traditional craft. Hence, there is a need for more research on the spaces of everyday practices in the informal settlements of the global South and how existing spatial practices can inform decision-making for future policy formulations. This proposition also adds to a better understanding of the role of place for the formation and adaptation of India's valuable household industries threatened by urban transformations.

Further, as much as the situated nature of the practices, the spaces have been shaped and continue to be shaped by local and global flows of capital and goods and associated networks of actors. Despite the informal setting, these relational and multilayered networks stretching near and far have shaped this thriving industry. The spaces of production are constantly in the process of being made and remade each year to prepare, store, and distribute idols. The multiplicity and flexibility of the spaces of production and the coexisting practices that shape these spaces are imperative for framing this book. This research argues that the physical spaces of production are shaped around the practices performed within them. Despite being in an informal setting, the industry has been thriving and growing the network of stakeholders beyond the existing geographies. This book contends existing literature on informality from the South; challenges undifferentiated perceptions of commercial, industrial, and productive functions within slums; and calls for studying intersections within informality. It reveals that in the case of traditional practices such as idol-making, the neighbourhood transcends aspects of informality to have become a site of cultural production, a location of a thriving industry, and a lived space. This is timely research of documenting and understanding this community's traditional spatial practices and places of production. Probably, in a decade or so, the emerging actors might transform the places of production so much that this study would stand as the missing link between traditional practices and spectacular contemporary exhibitions.

Second, I have adopted an architectural technique to map the activities on a spatio-temporal frame and propose this as a method to study the relationship between spatiality and material practices. The propositions of the practice theory method blog (Brown 2014; Shove 2017) and the freedom to experiment with the methods allowed room for introducing innovative methods to study practices. A combination of participatory, ethnographic, and architectural methods was used to record the activities of practitioners within idol-production spaces. The spatio-temporal analysis through this methodological innovation in studying practices based on spaces of production illustrates how the same spaces are shaped by changing practices over time. This methodology produced interesting perspectives into studying spatiality and their relationship with everyday practices. Further, it allowed me to study the perspectives of practitioners, and hence it adds to taking forward the understanding of theories of social practices and place-making.

I have outlined the key findings of the place-attachment study to articulate the need for understanding spaces within the neighbourhood. A key research outcome is a better understanding of the role of place for the formation and adaptation of valuable household cultural industries under urban transformation. Through the chapters, this book accounts for the place-based network of actors; spatial practices; and social, economic, and political dynamics of an inner-city neighbourhood largely dependent on their creative capital. Multiple actors have produced images and associated meanings to their everyday spaces. The participatory photo-study was particularly important in associating these multiple identities with the relational spaces of production and creativity. Informality, in a way, provides the flexibility and fluidity of the production of urban spaces with uneven regulatory and legal frameworks. A holistic analysis of the industry's operations through the exploration of social, spatial, and cultural practices and the relevant processes and flows produced a nuanced understanding of the political economy of the idol-making industry.

Finally, this book examines whether public policies are designed to accommodate traditional practices of household industries in marginalised urban settlements and their subsequent transformation due to consumer demands and global markets. In other words, it set out to examine whether or not the policies are inclusive of the everyday practices of 'urban dwellers' of 'slum' neighbourhoods in cities of the global South. The Kumartuli neighbourhood is a prototype for a much larger city-wide policy consideration with regard to informality and cultural tourism. I argue that although the city-government identified Durga Puja as a positive economic driver for boosting tourism and city-branding, enough concerns were not reflected within the same policy area to incorporate the idol-making residential neighbourhood and address the residents' everyday infrastructural challenges. However, there has been consistent large-scale displacement and subsequent commercialisation of several residences as a result of increased economic activity, the new tourism policy, and the earlier resisted redevelopment project. Also, the participatory data reflecting on the agency, legitimacy, voices of resistance, and power relations support the findings on disused and (mis)placed resources of the inner-city location. Hence, it would be fair to say that inner-city informal neighbourhoods such as Kumartuli, although consistently failing to be incorporated into mainstream urban planning consideration, continue to be otherwise impacted by wider public policies. In fact, this observation adds to ongoing debates on arts-led regeneration as a case from the South (López-Morales 2015) and stresses the superimposition of a dominant culture-led policy within an existing social structure. In doing so, this policy dispossesses and spatially shifts many to incorporate more dominant production functions thriving on existing resources.

This research acknowledges the ways in which multiple space–times are inscribed into what Doreen Massey terms a city's 'power geometry' (Massey 1993). This recognises that the freedom to extend one's actions in time and space is a form of power over space, time, social processes, and people, a recognition central to understanding contemporary cities (Graham and Healey 1999). The lack of investment or infrastructural degradation has already been leading to multi-scalar displacement, with some more successful artists moving out some of the storage and production functions away from Kumartuli, while others are moving residences on the basis of individual agency. Also, based on earlier discussions, it is not surprising that a few elite artists control much of the present idol-making market.

Additionally, the participatory data reflecting on the agency, legitimacy, voices of resistance, and power relations supports the findings on (mis)using the place-based resources and the importance of the inner-city location. This book initiates a discussion and opens up questions about the power structure and position to continue to carry on business in an informal setting. Empirical results show the spatial and economic reconfigurations within the neighbourhood, driven by the changing pattern of consumerism and rising global demands towards this traditional craft. The uneven power structure, stalled intervention, and politics resulted in temporary loss of livelihoods for some and doubling up of accommodation and storage spaces for a few others. However, the people most affected by the redevelopment plan have chosen to reorganise within the neighbourhood and retain workshop spaces. The resulting commercialisation of the neighbourhood risks further unplanned development utilising informality for unwanted gains while building on the existing networks.

It is much more than only the emotional bonding of the residents within the neighbourhood that adds to the place value: while Kumartuli has earned and situated place branding value, it has been vital to analyse the multi-layered network that operates to continue the industry in place. Representation of Kumartuli as the idol-making cluster of Kolkata and the primary brand that the place offers draws on relational perspectives of the imagination of the place by the residents. Place-attachment and the understanding of spatial, social and cultural processes, and economic functions inform an understanding of the industry's political economy. Despite the place attachment, displacement is a constant threat. Resistance to gentrification and arts-led regeneration are gaining momentum in the West (López-Morales 2015; Lees et al. 2018); research into placemaking and attachment from the South, such as this, adds to debates on marginalisation and the subsequent shifting of communities.

However, considering the wider implications of state tourism and cultural-industry policies, it would be fair to say that there remains mutual exploitation of sentiments relating to the religious and cultural festivals and the spectacular contemporary consumerism reshaping idol-making practices in Kolkata. In conclusion, I argue for a more nuanced understanding of the material spaces and infrastructure in addition to the practices performed in them to eliminate the effects of spatial and economic marginalisation of certain areas or neighbourhoods within a city. There remain questions for future explorations, whether there is a connection between the rootedness, embeddedness, and intangible cultural heritage. Also, there remains a concern as to how the inclusion of Durga Puja in UNESCO's ICH list would affect the already marginalised community of Kumartuli. To reflect on the enlistment of Kolkata's Durga Puja for the UNESCO world heritage, does it speak of more sustainable practices and how it might change in the future?

Recommendations

Through this research, I propose here a set of recommendations on the basis of my analyses and observations. Incremental development of the self-built housing stock could be a possible solution to address the informality that mitigates several challenges faced by the state-led formalisation instruments (Dovey 2013; Bhan 2019)—concurring with the recommendations in Bhan (Bhan 2019). I take a step further and suggest incremental development as one of the key policies to combat space shortages. As suggested in literature from the South, incremental housing and repairing and consolidation are vital in solving housing and accommodation problems. While incrementalism has been associated with housing and tenure in informal areas, it could be better used as a tool to provide housing solutions through local knowledge and networks in relation to building materials and labour (Van Noorloos et al. 2020). In doing so, a better understanding of local building practices and mapping the embedded communities is essential. This would enable opportunities around the availability of materials and affordability of the construction costs with more nuanced views of different actors within the self-building sectors. The Indian state instruments for formalising and providing for the poor through Rajiv Awas Yojana (RAY) or Ministry of Housing and Urban Poverty Alleviation (MoHUPA) guidelines seldom have provision for the well-being of the poor (Bardhan et al. 2018). As discussed in Chapter 5, on the provision of space for housing for the poor, these guidelines give approximately 25 square metres of space to each family unit. These guidelines do very little to solve the housing crisis, while officials as well have been quoted saying such provisions are inadequate (as discussed in Chapter 5). I argue that while these arrangements look inadequate, these are also not reflective of the everyday social and cultural practices of the communities inhabiting in them (Bhan 2017) and do not consider the production of sociocultural spaces. Since the provider paradigm of the state, such as in India, is linked with excessive quantification in housing provision for the poor, there remains a consumerist-oriented approach while neglecting the needs and well-being of the poor (Bardhan et al. 2018). A paradigm shift from provider to supporter policy at the state level is needed to enable incrementalism and public sector investment. It is imperative that we learn from community engagements and focus on localised needs while also understanding the broader flows within the cities to develop more consolidated solutions of scalability (Van Noorloos et al. 2020) as stressed during the deliberative workshop.

As an architect and urban planner trained in India, I reflect on the disconnect between theory and practice and understand why and how the subaltern voices of the slum-dwellers are never heard in the redevelopment/resettlement plans made with top-down Euclidian approaches by architects and planners. In architecture and planning schools, one such top-down problem-solving approach for a redevelopment project is to learn to structurally densify and rebuild to accommodate more residents and quantitatively increase units of residences. While I understand the need for learning innovative space-saving, problem-solving approaches for developing affordable housing, I also think that it is implicit to address local practices and understand in situ consolidation of the existing stock. Perhaps architecture and planning education need better communication from localised practitioners and learn from empirics around the world.

We need architectural and planning education in India to address this issue and identify alternative praxes of dwelling and working in these marginalised *basti* neighbourhoods where auto-construction is the most prevalent building mode. To widen and steadily include such methods of construction knowledge of localised building materials, construction techniques are essential. Also, an understanding of the citywide complex relationship of labour, materials, and financing is required to consolidate the informal housing stock. Possibly, there is also scope for community participation to produce more nuanced knowledge of practices and livelihoods to facilitate better housing policies. Here, I stress the relevance of looking from the place and learning the needs of people facing housing inadequacies instead of unified top-down nationwide policies on housing shortages.

Although my research questions were studying the transformation of the Kumartuli cluster, using a Southern theory lens, I have reflected on the spatial and social changes that affect Kumartuli's environmental and economic functions. Addressing calls for critical analyses of the political economy of urban informality (Banks et al. 2020), I analysed and discussed in-depth the multi-layered and situated practice of idol-making along with its wider socio-economic and spatial networks. I discussed why spaces are contested and constantly give way for more productive functions to be performed in them as, due to the real estate values and existing place-based resources and networks, they continue to be associated with idol-making practices. However, the transformations and the failed planned intervention from the previous government have led to the worsening of dwelling conditions. The reasons for the failure of this plan have been discussed in the previous chapters; however, based on the interviews with stakeholders, specific steps could have been taken to have better results from the plan. While mainstream planning practices exclude informality, a more holistic approach to understanding the need and everyday realities of slum areas is required. As suggested in this research and elsewhere, politics at the neighbourhood level involves several intermediaries who practically necessitate a dialogue between the local political leaders and the residents. It is perhaps essential to involve their participation in the planning process and involve communities on a larger scale much earlier in the process to discuss and draw on dwellings and existing practices. While it might sometimes be challenging to adopt successful policy formulation from Western or other Southern cities, learning from less successful case studies is crucial. Maybe the solution does not involve rebuilding but incrementally repairing and consolidating what already exists. I echo Bhan (Bhan 2019) in stressing the repairing and consolidation of existing resources and infrastructure, on both horizontal and vertical scales utilising community engagements, organisations, and perhaps public sector utilities and investments. As planners suggested during interviews, seeking local knowledge and enhancing communication between stakeholders are crucial. Hence, the role of the planner also requires listening and understanding of these needs and to devise policy recommendations on the basis of these nuanced views of communities.

In the case of Kumartuli, residents have specific space requirements to perform the creative and productive functions of the idol-making practice. These requirements can only be reflected through localised knowledge of the practice and the seasonal influx. As directly indicated several times throughout this research, particularly during the deliberative workshop in the aftermath of the COVID-19 pandemic, placing standardised housing and an in situ redevelopment plan would not be appropriate in the case of Kumartuli. As planners suggested that there were dialogues among themselves, the community, and the political representatives, still an acceptable solution could not be reached. The planners remained constrained by the quantitative budget and tickboxing the redevelopment agenda of the local government, while the residents felt that the planned residential and production units would be insufficient. First, the discussions between the different stakeholders, which happened quite late to solve the resistance amicably, needed to happen in the beginning. Also, I believe that the redevelopment plan would be better suited as a support plan that would provide resources, public sector utilities, and financing opportunities for the residents while they continued to live and work from their existing spaces: a view shared by the community representatives in the deliberative workshop.

Moreover, the other concern arising from this research's results is to tackle unintended marginalisation and spatially shifting residents to make space for commercial purposes. Informality provides the scope for avoiding regulations that have been misused by developers for illegal gains. It is essential, therefore, for the Kolkata Municipal Corporation to be cautious of the implications and end uses of these developments and continue to reserve the rights of the age-old tenants of Kumartuli.

Personal reflections

Towards the end of my PhD in 2020, Durga Puja gained an international audience by submitting an appeal for nomination in UNESCO's ICH list. However, there was an ongoing change in the perception of Durga Puja and how it is celebrated in the city. While in the past, the general perception in Kolkata was that *kumars* of Kumartuli sculpted the most iconic idols; however, my fieldwork suggested a new and emerging group of actors that are shifting the vital production functions elsewhere. I was surprised to discover the scale of corporate investments in sponsoring the big-budget neighbourhood pujas. Interviews suggested the changing landscape of idol and *pandal* construction and distribution. Due to the limited scope of this research, I have not been able to investigate the full range of new stakeholders in detail, but I have been able to present the changing scenario from both producer and consumer perspectives. Moreover, it made me realise the heterogeneity and the differential that have been growing between the caste-based artists of Kumartuli and the emergent group of sculptors engaged in idol-crafting. While the caste-based artists of Kumartuli traditionally sold only idols, the new class of artists also offer ambient concepts of presenting the idol and *pandal* to suit the competitive market of winning accolades and sponsorships to the neighbourhood festivities.

I started my fieldwork in 2017 around the Durga Puja celebration that year. I documented the activities of Kumartuli, relived my old memories of Durga Puja in the city, and witnessed the grandeur of the state-sponsored immersion parade. Following the autumnal peak in Kumartuli, I carried out my fieldwork. I interviewed a number of government officials, local politicians, and middlemen to find answers to the growing discontent to the failed plan and the subsequent displacement of certain idol-making functions. Interviews revealed the importance and momentum of the KMDA plan and the deep-rooted split effect it had on the residents. The conversations I had with planning officials and local representatives painted distinctively different pictures of the resistance struggle.

Nevertheless, I was amazed by the resilience of the community. While Kumartuli is still reorganising after the failed attempt at formalisation, the Durga Puja celebration has been taken over by corporate players elsewhere. Interviews suggest that to remain in business and to save their livelihoods, Kumartuli residents had to take actions fast and reorganise their spaces. However, the complexity of tenure and the winners and losers triggered by the plan made me question the proposal's effectiveness and investigate further the housing sector provider paradigm of the state, which was initially not a goal of this research. This insight soon highlighted the increasingly tricky situation the planners were positioned to oblige in incorporating the housing guidelines into the existing practices of Kumartuli.

About a year later, I found out from newspaper reports about the plans for the nomination. After completing my doctoral research, I secured the ESRC postdoctoral fellowship to complete a further round of investigation and ask questions about the nomination of Kolkata's Durga Puja and how that changes or alters the perception of Kumartuli to the various stakeholders I interviewed before. However, the COVID-19 pandemic beginning in 2020 had resulted in a shift in how people live, work, and even conduct research globally. Not only did the pandemic reinvigorate the disjointed public services, but also public health, safety, and hygiene were key concerns in these densely populated inner-city areas across the global South. Due to limitations to conducting face-to-face overseas interviews, I conducted a deliberative workshop with a representative list of a limited number of participants. Crucially, however, despite the limitations, the key discussions of this deliberative workshop reiterated my earlier research findings. While there has been academic interest within the disciplinary boundaries of architecture, planning, and anthropology in Kumartuli for many years, I believe a more nuanced understanding of the relations of idol-crafting and allied industries, which has been rooted in the neighbourhood and elsewhere in the inner areas of Kolkata, is crucial. Further, more in-depth investigations of the political economy of the informal sector and cultural industries that form part of the unorganised sectors are essential.

Notes

1 For further reading on contemporary Durga Puja installations and crafting, see Guha-Thakurta, T. (2015). *In the Name of the Goddess: The Durga Pujas of Contemporary Kolkata*. New Delhi, Primus Books.
2 For more in-depth discussions on perceptions of modernity and the colonial influences on the culture and the built environment of inner-city Kolkata, see Chattopadhyay, S. (2005). *Representing Calcutta: Modernity, Nationalism and the Colonial Uncanny*. Oxford, Routledge. **2**.

References

Banks, N., D. Mitlin and M. Lombard (2020). Urban informality as a site of critical analysis. *The Journal of Development Studies*. **56**: 223–238.
Bardhan, R., R. Debnath, J. Malik and A. Sarkar (2018). Low-income housing layouts under socio-architectural complexities: A parametric study for sustainable slum rehabilitation. *Sustainable Cities and Society*. **41**: 126–138.
Bhan, G. (2017). From the *basti* to the 'house': Socio-spatial readings of housing policy in India. *Current Sociology*. **65**: 587–602.
Bhan, G. (2019). Notes on a Southern urban practice. *Environment and Urbanization*. **31**: 639–654.
British Council (2019). Mapping the creative economy around Durga Puja 2019. British Council.
Brown, K. (2014). Global environmental change I: A social turn for resilience? *Progress in Human Geography*. **38**: 107–117.
Chatterjee, P. (2011). *Lineages of Political Society: Studies in Postcolonial Democracy (Cultures of History)*. New York, Columbia University Press.
Chattopadhyay, S. (2005). *Representing Calcutta: Modernity, Nationalism and the Colonial Uncanny*. Oxford, Routledge. **2**.
Dovey, K. (2013). Informalising architecture: The challenge of informal settlements. *Architectural Design*. **83**: 82–89.
Graham, S. and P. Healey (1999). Relational concepts of space and place: Issues for planning theory and practice. *European Planning Studies*. **7**: 63–646.
Guha-Thakurta, T. (2015). *In the Name of the Goddess: The Durga Pujas of Contemporary Kolkata*. New Delhi, Primus Books.
Heierstad, G. (2017). *Caste, Entrepreneurship and the Illusions of Tradition: Branding the Potters of Kolkata*. London, Anthem Press.
Lees, L., S. Annunziata and C. Rivas-Alonso (2018). Resisting planetary gentrification: The value of survivability in the fight to stay put. *Annals of the American Association of Geographers*. **108**: 346–355.

López-Morales, E. (2015). Gentrification in the global South. *City.* **19**: 564–573.

Massey, D. (1993). Power-geometry and a progressive sense of place. *Mapping the Futures.* London and New York, Routledge: 59–69.

McFarlane, C. (2012). Rethinking informality: Politics, crisis, and the city. *Planning Theory and Practice.* **13**: 89–108.

McFarlane, C. and J. Silver (2017). Navigating the city: Dialectics of everyday urbanism. *Transactions of the Institute of British Geographers.* **42**: 458–471.

Millington, N. (2019). Critical spatial practices of repair. *Society and Space.* Retrieved June 7, 2023, from https://www.societyandspace.org/articles/critical-spatial-practices-of-repair.

Roy, A. (2009). The 21st-century metropolis: New geographies of theory. *Regional Studies.* **43**(6): 819–830.

Shekhar, S. and S. Shekhar (2021). Slum development programs—An overview. *Slum Development in India: A Study of Slums in Kalaburagi.* Springer International Publishing: 135–157.

Shove, E. (2017). *Practice Theory Methodologies Do Not Exist—Practice Theory Methodologies.* Retrieved July 16, 2023, from https://practicetheorymethodologies.wordpress.com/2017/02/15/elizabeth-shove-practice-theory-methodologies-do-not-exist/.

Simone, A. (2004). People as infrastructure: Intersecting fragments in Johannesburg. *Public Culture.* **16**(3): 407–429.

Van Noorloos, F., L.R. Cirolia, A. Friendly, S. Jukur, S. Schramm, G. Steel and L. Valenzuela (2020). Incremental housing as a node for intersecting flows of city-making: Rethinking the housing shortage in the global South. *Environment and Urbanization.* **32**: 37–54.

Methodological appendix: research strategies

As briefly stated in the introductory chapter, several inner wards of India's older cities are inhabited by poorer communities involved in creative livelihoods. These neighbourhoods are generally informally built, well connected to the city's networks, and have been through legitimisation processes. This case study reflects on a similar neighbourhood and draws on citywide and national politics while portraying socio-economic inequalities and rapid urban transformation. The central research question of this study is to understand the relationships among material processes, everyday social practices, and spatiality through phenomenological explorations of place-based functions. The aim of the research was the context- and place-specific—to trace material processes and everyday social practices that shape the culturally embedded and situated idol-making practices in Kumartuli. It also assessed the reconfiguration of the spaces of production within the place-based network of idol-making and formed the basis of the research that explores the link between spatiality and idol-making practice in the said neighbourhood of Kolkata. In the wider context, the case study explores a cultural, traditional industry based in an informal neighbourhood of a postcolonial city. To answer the research questions, I designed a combination of an innovative method combining human geography and architectural methods, which I have described in detail in the next sections. The research also included three phases of fieldwork in 2017, 2018, and 2022.

Practice theory research methods

Practice theory being the central concept of this research, the inquiry into culturally embedded practices of idol-making framed the ethnography-inspired approach followed in this research. As Shove (2017) puts it,

> [T]aking 'practice' as a central conceptual unit of enquiry generates a range of distinctive questions. The choice of methods depends on which of these questions you want to take up and pursue. Using practice theory is thus not directly tied to certain methods, but the choice of methods is—as always— dependent upon your specific research question.
>
> (Shove 2017)

Shove also adds that theories depend on how problems are defined or how lines of enquiry are formulated, giving enough scope to experiment. Adding that there is no simple link from theory to the method, she opens up the scope for formulating new lines of methodological techniques to work with, which are in relation to the particular research question and designed to suit the requirements of the study. The practice theory methodologies' blog enlists a set of seven intriguing propositions to encourage further debates and the development of the practice theory methodologies (Shove 2017). An interesting proposition, number (4) states: '*Inventive and multiple methods, units, samples, etc. are particularly useful for exploring practices at different scales, in relation to changing social patterns and variably interconnected actors, dictates the methods adopted*'. Also, other relevant propositions (2), (3), (6), and (7) have been useful in framing this research. Crucially, as Mylan and Southerton (2017) stress that these are particularly dependent questions on what spatial and temporal scales are the 'coordination' of practices occurring at. They also highlight what spatial, temporal, or social interaction questions we are seeking to answer through the methodological techniques of the research. The flexibility of the practice theory framework is that it does not prescribe methodology. Therefore, it is important to address the central practice and the practices that happen in coordination with it through an innovative mix of methods tailored to get the desired answers to the relevant research questions.

The research methods used in this study might not have been singularly used earlier to study practices solely, but most of the methods are used to study everyday lives, livelihoods, and practices of communities. I have chosen to use specific methods in conjunction with others to fit the purpose of this research. In this case, the research objectives largely governed the research methods. The following sections describe the research methods adopted, adapted, and the underpinning research objective associated with them.

A combination of ethnographic research methods

Ethnographic approaches informed this research, as it seeks to trace the everyday practices and activities through participation in varying degrees by the researcher (Herbert 2000). The methodological approach in ethnographic research, which involves tracing objects, products, or people located around multiple geographical locations transnationally, forms the basis of multi-sited ethnography (Marcus 1995). A product's supply chain mapped through global consumerism is an emerging trend in interdisciplinary studies, including media, human geography, science, and technology studies. Cook (Cook 2004), in response to Harvey (Harvey 1990), traced the market trend for the production and distribution of common fruit, papaya, to make connections between the producers and their distant 'western' consumers. This type of ethnographic study involves research carried out in multiple locations and addresses geographies of production, transportation, and consumption of commodities and the underlying narratives of exploitation of the labour market (Cook 2004).

Another example of a visually compelling multi-sited ethnography was carried out to trace the mass production of flip-flops. A flip-flop—a globally used product—shows the wide geographical reaches of the production process (Knowles 2014). This study connects the production process of flip-flops from its inception to disposal while mapping all the actors and their physically distant locations and linkages. This visual documentation and mapping inspire this research design. This mapping exercise across multiple locations is particularly useful in the context of the idol-making industry. I have used this technique to study the spaces and places of cultural production within informal urban settlements involving various networks of practices.

In order to map the extent of the idol-making neighbourhood and its wider socio-spatial network, the ethnographic approach was modified to visually map and document the supply chain of the idol-making industry across Southern parts of Bengal. Spatially mapping locational information through identifying sets of actors and drawing layers of networks for presenting the collected information was adopted as the key documentation technique used in this ethnographic approach. Studying the idol-making practice opened up connections and competition among the practitioners, suppliers, and allied business owners within the idol-making industry. In order to trace some of the linkages between materials and labour, I visited areas like Krishnanagar and Chandannagar, where allied idol-making practices are performed. Following the materiality of the idol includes tracing the procurement of raw materials from different locations and the production of ancillary objects for the finished product documented through interviews with stakeholders situated in Kumartuli and Greater Kolkata. The idol is traded locally as well as shipped globally. The main items required for the production are straw, clay, and bamboo. The different ancillary components used in the finished product are connected with the idol-making industry and are situated around Kumartuli. The media, the press, and anthropological literature on potters cover the two ends of the process—the idol's production and the festival's celebration (Bean 2011; Ghosh 2012; Heierstad 2017), but the linkage of the supply chain is underrepresented in literature.

Additionally, photography has been used in mapping the neighbourhood and other sites. The idea of using photography was adopted carefully from visual sociologists' and anthropologists' accounts of avoiding the 'sample bias' in researcher-produced data. In achieving this, the photographs were '*taken without regard to particular content or aesthetics*' (Krase 2012). The frames of the photographs were selected on the basis of systematically mapping the entire neighbourhood and its surroundings in detail. Photography was a crucial aspect of this study in visually documenting and recording temporal and seasonal variations of the idol-making practice and the spaces and places in which these practices are performed.

To examine the public spaces and the individual production units of idol-crafting, in-depth documentation of the various production spaces coexisting in the idol-making cluster's living spaces was carried out. Additionally, this spatial arrangement was analysed to assess the importance of place for idol-making as a cultural and spatial practice. In order to carry out the study of spaces, an architectural technique of following user activities over a spatial and temporal frame was useful. For this, in addition to visually documenting the neighbourhood through photography, an architectural technique of mapping through line-drawing of buildings and flows and movements within these, inspired by architects Sarah Wig-glesworth and Jeremy Till's writings, was adopted (Wigglesworth and Till 1998). The dining table experiment formed the basis for conceptualising coexisting living and working spaces and how these were oriented to function effectively. In order to design their living and working spaces in their residence, they used the dining table experiment to figure out the use of space. They found it impossible to dis-tinguish spaces even with grading or separating the space. This was an interesting experiment not only to analyse the spatial practices but also because the technique of mapping used in this experiment was designed to visualise the users' activities on a particular space over a particular practice of dining. Interestingly, the dining table experiment produced a chaotic array of lines mapping the users' hand move-ments over the course of dinning for a meal. These movement diagrams, along with the spatial alignment of other furniture, the dining table and the wall in the living (working) space, symbolise the coexistence of the public and the private lives of the residents.

Here, I enquire what these movements mean when drawn as a line on a piece of paper. Whether or not this analytical tool, inspired by the dining table experiment, can be used to study the line drawings of paths taken by users of a particular space to understand the relationship between practices and spatiality appropriately is a question. Following these thoughts, I examine whether such traces of everyday practices of human beings left behind in space through their movements could be compared to the reductive traces, which leave a long-lasting effect on the spatial structure and the subsequent codes of practice. I also seek to understand whether the interlinked nexuses of practice have underlying invisible lines of human move-ments bound by certain hierarchical codes of human interactions, more so in spaces of coexisting practices such as living and working. In the case of Kumartuli's spaces, I study and trace the movements of users of particular spaces over a period of time to determine these answers. One such line drawing during my fieldwork is presented in Figure A.1. The spatial implications of these drawings are explained in detail in Chapters 3 and 6.

Participatory visual research methods

Participatory methods are increasingly being taken up in postcolonial and feminist critical perspective research to facilitate collaborative and dialogical knowledge production. A participatory approach was needed to explore the place through resi-dents' lived experiences. Based on the required information and research design, participatory methodologies range from focus group discussions to action research workshops, which are representative of a community. Participatory methodologies are widely used in low-income groups to identify their problems and vulnerabili-ties and assess the type of required interventions from the participants' perspective (Moser and Stein 2011). Methods of participatory research have inspired research processes which are less extractive and more participatory and question imbal-ance within fieldwork and academic writing in ethnographic research (Gubrium and Harper 2016). Inspired by the success of participatory approaches, I started my fieldwork with plans of detailed interviews to address issues around vulnerability and displacement to understand the resident's perspectives.

The issues that came forward through the interviews were grievances, such as the quality of life, housing rent sharing, and unaffordability outside Kumartuli. Social vulnerability and displacement fear exist due to the precarious ownership pattern and land tenure in the slums of Kolkata. However, the participants' consciousness of being recorded while interviewing was an obstacle. Participants spoke less and were constantly worried about being overheard by neighbours, passers-by, and mostly the local political middlemen, although they were assured of privacy and anonymity. I took field notes from the initial phase of the fieldwork wherever possible to record these difficulties. However, I felt that the goal of knowledge production in a more unbiased way was not achieved. This interview approach, particularly important to understand the community's views towards the ongoing transformation, was not coming through. Interviewees were constantly hesitant to express their grievances and challenges faced due to poor infrastructural conditions and the situation of land tenure and housing since the failed redevelopment process. I later took a different approach based on these observations and modified the participatory methods. The asset-based approach, combined with the participants' photography study, was used in the following fieldwork phase. The challenges faced during the earlier interviews were reasons for carrying out participatory photography and subsequent photo-elicitation interviews (Harper 2002). These combined methods form the basis of understanding the participant's perception of their neighbourhood and remedial opinions as representative of a community.

Photographic methods can be used as a tool for social research to provide meaningful and grounded analysis of acquired information (Schwartz 1989). Knowledge production in social research aims at the social construction of knowledge, informing the social practices and participants in the research about social and economic change and producing valid contextual knowledge centred on the subjects (Lombard 2013). The construction of participatory knowledge through participatory visual research, starting from the early 1970s, is very much similar to telling stories of the participants' lives and livelihoods. The participants describe the places and spaces that they use and map them as part of their life process in this kind of approach (Elder 2001). The use of visual methods is becoming increasingly common in geographical research. Images are part of the discursive production of geographical knowledge. Auto photography offers ways of understanding these images captured, specifically, how marginalised communities understand space and place. Residents' perspectives of place meaning and how the place offers much influence on the type of industries and livelihoods and the communities involved in can be explored through auto photography.

A participatory visual methodology is an emerging technique social scientists and anthropologists use for ethnographic research (Gubrium and Harper 2016). Prosser and Loxely (Prosser and Loxely 2008) note that different scholars have used visual ethnography and participant photography or videography as part of collaborative knowledge production by literally '*placing cameras in the hands of participants*' (p. 18). Various disciplines such as sociology, human geography, media studies, and the like have a growing number of researches that are placing impartial relations to generating knowledge and distributing power between researchers and the participants. Some of the more notable works are by Wang and Burris (Wang and Burris 1994) visual method, Photovoice by Margolis (Margolis 1988; Margolis 1994) and more recently by Milne (Milne 2012; Wilson and Milne 2016).

Informed by postcolonial theory, the dual meanings of informal identities help understand the people's perception of the spaces and the place they live in. The use of auto-photography (participatory visual method) in marginalised/informal areas where the interviews are not possible 'as a way of accessing residents' perspectives of place meaning', as part of qualitative research methods, enables to understand participants' perceptual observations that may be limited through interviews or more conventional methods (Lombard 2013). Using participatory photography and photo-elicitation within a community, facing an urban transformation as a means of gathering data can help answer relevant research questions on social organisation and provide evidence of the place-based notion of the participants. Participants often have memories and perceptions of the place, which are shaped and reshaped due to the ongoing changes in their surroundings. The participants were able to evoke a deeper meaning of their surroundings from the photographs they took. The participatory visual study was used to assess the sense of place and the importance of the location of the industry to the community.

Analytical aspects of using auto-photography give a psychological exploration of what the place means to the respondents and their self-identities and how they identify with the place. Also, multiple meanings and emotional significance of a place for the users, passers-by, or residents and the reasons for including certain places instead of others by the respondents make the technique quite useful for bringing forward psychological insights into the people's perception of the place.

The negative aspects of this technique lie in the lack of interpretation and only producing imagery taken by the participants. Therefore, a careful consideration in the interpretation of the imagery and follow-up interviews with the participants to collectively generate and record the meanings while keeping the anonymity of the participant and complete ethical consideration help in building up the best possible auto-photography method. This results in a socio-spatial construction of a place, negating the duality of understanding the urban setting of the global South representing both ordinary and extraordinary spaces through participants' visual imagery and narratives. The participants' observation and sense of place informed by the theories of the subaltern are hence recorded, making this approach a rather interesting one to social scientists and urban theorists.

In this scenario, the respondents, rather the marginalised communities, have a voice to speak of the places they use and how they interact with these spaces daily. The respondents have articulated images of their own production to contribute to common knowledge, enabling me to construct a narrative of relationships on the basis of their emotional reflections. This technique raises ethical and safety issues for the participant, and I have been careful regarding the anonymity of the participant and considered all challenges against exploitation. The deprivation and unethical means of research often are discriminative. The community are sometimes unaware of the publication of research. The fact that they were consented before being photographed and their photographs to be published made them feel reassured. I have provided information about the research and assured them of anonymity. Also, if participants are willing to learn more about the publication of this research, I have provided them with my contact information.

During the photo-study, I put questions like '*What sort of photo would you like to send to your distant consumers who are unable to visit you and usually place orders over the telephone or mobile apps*'. In response, participants have photographed, for example, the riverfront, their workshops, dilapidating roofs, workshop fronts, etc. They have photographed their friends and neighbours, whom they say are their 'lifelines'. Participants were also asked what sort of remedy or governmental intervention they think would be adequate to alleviate their degrading environment.

Fieldwork

Fieldwork has been the basis of this research to engage with the generation and analysis of empirical data critically. The fieldwork was conducted in two phases between September 2017 and May 2018, and then a final round of investigation was conducted in February 2022.

Phase 1: September to December 2017

I conducted a participant observation study to initially document the production process through the 'follow the idol' method. In order to study the site at the core of this ethnographic fieldwork in Kolkata, I selected the busiest season of the year to start the participant observation or 'being there' (Marcus 1995; Hannerz 2003). Physically I had been there before as part of an earlier study group, but never so during this time of the year. During the preparatory season of Durga Puja, Kumartuli is overcrowded with many people; the industry engages directly and indirectly. My first visit to the site was days before the annual celebration of Durga Puja of 2017.

During the initial phase of the fieldwork, from five participating *kumars*, I could trace 30 respondents within their network of suppliers and relatives in other locations. Based on these initial interviews, mapping the supply chain network in South Bengal and many consumers beyond it was possible. The artists and their families were the primary sources of the collected data, and, through their contacts, the network opened up. Hence, five such *kumars* and their immediate family were interviewed first, from the younger generation to middle-aged and older artists. The range of the interviewees was kept as broad as possible to get better analytical results. This method carefully incorporated novice trainee *kumars* to well-established (famous) artists. Also, as women of the household are involved in the family businesses, a few interviews with female artists were conducted to get a gender perspective. Children under the age of 16 were not considered in this study. Additionally, the seasonally hired workers were not available during the initial phase of fieldwork; hence, they were interviewed in the subsequent phase. Formal and informal suppliers of each item used in the production were contacted. Some of them were unable to participate as it was during the off-season, but the operational dynamics of their businesses were recorded from the idol-makers' perspective. I interviewed them during the second phase of the fieldwork. In total, I interviewed 42 respondents (detailed in Table A.1).

Through the ethnographic approach inspired by 'follow the thing', following the idol posed many challenges. Hence, I followed the traders based in Kumartuli and nearby areas to map the locations of origin of each material of the idol-crafting

Table A.1 Table of interviewees

Role	Interviewee Profile	Date of Interview
Artist	Male, middle-aged, average artist	03.10.17
	Male, middle-aged, average artist	03.10.17
	Female, middle-aged, well-known artist	03.10.17
	Male, young, well-known trained artist at Baranagar	13.10.17
	Male, young, average artist	22.10.17
	Male, middle-aged, average artist	22.10.17
	Male, young, struggling artist	25.10.17
	Male, middle-aged, trained famous artist	28.10.17
	Male, old, famous artist, Bangladeshi origin	09.11.17
	Male, young, average artist	11.11.17
	Male, old, waning fame	20.11.17
	Male, young, struggling artist, politically active	20.11.17
	Male, middle-aged, locally famous at Chandannagar	22.11.17
	Male, middle-aged, average artist at Chandannagar	22.11.17
	Female, middle-aged, struggling artist at Chandannagar	22.11.17
	Male, middle-aged, famous for sculpting statuettes at Krishnanagar	25.11.17
	Male, young, known by his father's reputation at Krishnanagar	25.11.17
	Female, old, famous for her deceased husband's name	22.04.18
	Male, middle-aged, average artist, shared workshop	10.05.18
Seasonal workers	Male, young, from Krishnanagar	23.04.18
	Male, middle-aged, from Krishnanagar	23.04.18
	Male, middle-aged, from Ranaghat	25.04.18

Role	Interviewee Profile	Date of Interview
Supplier	*Shola* and *Dak* ornament shop in Kumartuli	09.11.17
	Bamboo storekeeper at Kumartuli	09.11.17
	Sells all ritual items	20.11.17
	Very small *shola* carving shop in Kumartuli	20.11.17
	Makes *shola* items from the family home at Krishnanagar	25.11.17
	Young, wholesale supplier of ornament	05.05.18
	Clay supplier, Kumartuli *ghat*	07.05.18
Co-op members	Secretary, cooperative of idol-makers, Kumartuli Mritshilpi Sanksritik Samiti	08.11.17
	Joint-Secretary, cooperative of idol-makers, Kumartuli Mritshilpi Sanksritik Samiti	09.11.17
Political	Councillor, Ward no. 8, KMC	09.11.17
Representative	Councillor, Ward no. 9, KMC; telephonic conversation	17.11.17
	MLA, Minister of State, contacted	10.11.17
	PA of the councillor of Ward no. 8, locally active	09.11.17
	Ex- Member of Assembly, Member of Parliament	20.11.17
Government	Ex-chief planner, KMDA	11.10.17
official	Assistant planner, KMDA	20.10.17
	Associate planner, KMDA	20.10.17
Client	Puja committee member	30.04.18
	Puja committee member	30.04.18
	Puja committee member	30.04.18
	Family festival organiser	12.11.17

process. The main items required for the production of idols are straw, clay, and bamboo. Straw being an agricultural by-product of paddy (widely cultivated in South Bengal) was extremely difficult to trace to a single location. Therefore, the information provided by suppliers and the available written literature in Bengali (Agnihotri 2017) were the only sources of mapping the location of straw production specific to the idol-making industry. Similarly, clay can be dug out from the river, which was the conventional practice. Due to government policy restrictions and safety measures, clay is now collected from around 90 km south of Kumartuli, towards the mouth of the river Hooghly. Again, there is no restriction in digging clay from the riverbed further down south. The wide range of locations posed a challenge in covering all the sites of clay collection. Therefore, I interviewed the merchants carrying boat loads of clay to Kumartuli. The supply chain map was drawn on the basis of information provided by these suppliers. Additionally, I also visually mapped and documented the neighbourhood. A detailed inventory of types of workshops based on size and the operational structure was drawn to study further the nitty-gritty of the everyday practices of the residents of Kumartuli. This inventory was important in determining the further mapping exercise within workshops during the second phase of the fieldwork.

Also, I interviewed the financers and the members of the artists' cooperative. Political leaders, municipal representatives at the City Corporation, and their neighbourhood-level middlemen were contacted; some responded, and a few refused to comment. The government officials, like planners and engineers involved in the execution of the proposed development project (KMDA 2009), were crucial informants for statistical data about the neighbourhood and the planning proposal. I interviewed them and collected data from the previously prepared survey reports. Semi-structured interviews with policymakers helped to understand their view on this phenomenon of redevelopment and rehabilitation. Other than them, a very small number of political representatives were not available to answer any questions about Kumartuli. Telephonic interviews were very brief. The planners from KMDA elaborated on the reports and their experience during the failed implementation procedure of the redevelopment. I collected many newspaper reports, articles, and policy documents to overcome the data gap from the political representatives. I conducted a few additional interviews with clients, local residents involved in worshipping the idols, and tourists who form an integral part of the network. The population being high, it is almost impossible to carry out in-depth interviews with even the smallest sample. A very few clients were considered as interviewees for explaining the value of the idol to a consumer and explaining whether they are aware of the functioning of the industry.

There might be expensive Durga idols prepared by famous artists, but most idols are a deity to worship for the local community, and affordability for the consumer is important. The demand for idols never ceases as idols are meant to be immersed after worship and rebuilt the following year. The artists of Kumartuli prepare not only heavily priced Durga idols but idols of other major festivals as well and struggle to sell them at a mutually agreeable price with the customers. To understand the customers' perception of a high-priced Durga idol, I interviewed a few organisers from community festivals in Kolkata. Perceptions of the idol for a consumer are that, whether they use it for a family puja or a big neighbourhood-sponsored festival, the idol always remains the means of worship, festivity, fun, and social gathering, and yet it must always be affordable. Perhaps the underlying exchanges that happen within the production network often remain unnoticed by the consumer.

Phase 2: March to May 2018

The artists and their families' living and working spaces within the same building were documented through participants' observation and mapping. Documenting the everyday practices within the workshops was inspired by Wigglesworth and Till's Dinning table experiment (Wigglesworth and Till 1998). I spent through entire days drawing lines of activities and usage of spaces within the workshop-residences of three artist-families. During these days, I sat down at corners of workshops or outside on the benches by the streets. I drew movement diagrams documenting every action of the users, that is the main artists and hired seasonal workers, of the production spaces over a period of time (Figure A.1). This mapping exercise produced linear movement patterns on a paper illustrating the use of space on a basic plan of the production studio residence. I used different coloured pencils to note the movement of different users and added handwritten notes to accompany them. For the purpose of this book, I have remade one such map where I replaced the coloured lines with a range of dotted and dashed lines to illustrate the same (see Figure A.1). During this study,

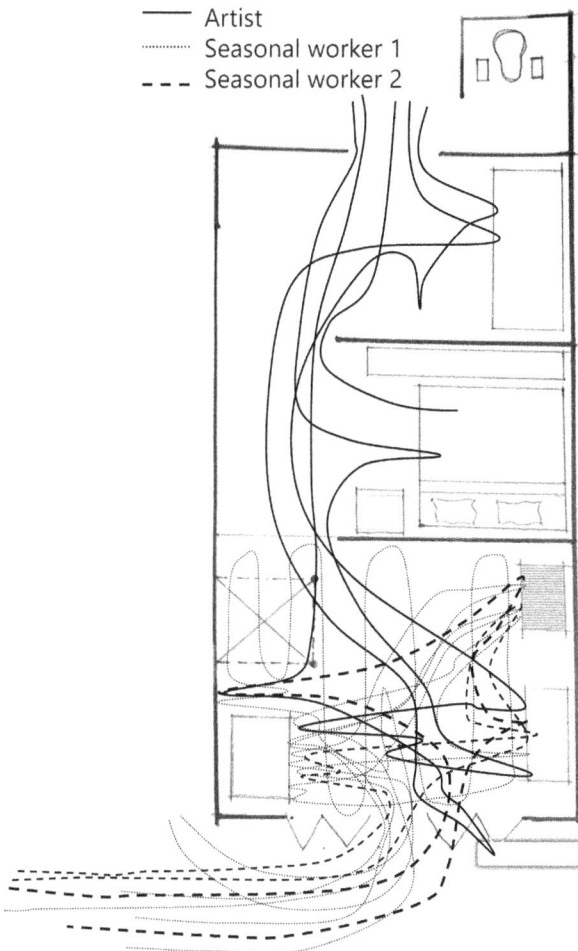

Figure A.1 Movement diagram of users within a workshop unit during a working day
Source: author

I took field notes alongside the mapping on the basis of informal conversations with the members of the workshops. This documentation records a snapshot of the practices performed in the production spaces over the span of a day during the preparatory phase of the idols.

Participatory visual methods were introduced in the research design to overcome the obstacle of previously perceived barriers in interviews. In this phase of the fieldwork, participants were selected through snowballing from professional contacts, friends, and earlier communications. I provided the participants with printed forms describing the entire photo-study and gave them a day's time to think about their involvement in the study. Afterwards, on their consent, participants were given disposable cameras to take photographs of the places in their neighbourhoods they relate to and of special or typical characteristics of the neighbourhood that they identify with. Also, they were asked to take photographs of their workshops and working and living conditions to illustrate their everyday lives. This was particularly useful in understanding the participants' attachment to their immediate surroundings. I distributed cameras to 10 participants and took these back after a few days, as detailed in Table A.2. An elderly artist, who felt uncomfortable to use the disposable camera, asked me to accompany him while taking pictures. Another woman used the help of her 17-year-old son to take pictures. This

Table A.2 Basic details of the photo-study participants

Resident	Profile
R1	Male, 45, artist, has limited clientele and income, has a larger workshop, and lives a few hundred metres away from the neighbourhood core
R2	Male, 17, high school student, preparing for board exams, lives with parents and sister in a one-bedroom accommodation
R3	Male, 49, artist, relocated with family to Kumartuli 2, currently rebuilding demolished workshop at Kumartuli
R4	Male, 19, local youth, does not belong to potter or *shola*-carving family. But was born and brought up in Kumartuli neighbourhood
R5	Male, 55, artist, lives near the riverfront and railway track and has an independent workshop next to this residence.
R6	Male, 42, artist with average acclaim, has a newly built house in the centre of the neighbourhood on the shared family land.
R7	Female, 40, housewife, lived in Kumartuli since childhood and currently resides with family on a nearby street, a few hundred metres away from her husband's workshop
R8	Male, 22, artist, works with his father and studied fine arts to enhance his technical skills, moved to Kumartuli 2 with his family and currently rebuilding their workshop in their old plot
R9	Male, 25, *shola*-carving tradesman, lives with his maternal uncle and family in Kumartuli since childhood.
R10	Male, 70, artist, famous in the 90s and 2000s, well-respected in Kumartuli, lives with his grown-up son and family in a two-room accommodation behind his workshop.

little modification in the method helped me get two different views of important participants who otherwise would have been unable to be a part of the study. After printing the photographs, some photos were immediately discarded due to a lack of light exposures and, hence, very low-quality prints. Photographs of children and their toys were returned to the participants, a decision I made considering the ethical boundaries. Follow-up interviews with photo elicitation to interpret and understand the photographs taken by the participants are of utmost importance. Their reasons for taking the photo and how they identify or relate to the photographs were recorded.

The place is perceived as a lived space from the perspective of the user, irrespective of the formal and informal nature of the land and building ownership within the neighbourhood. Applying a place-based approach to understanding practices within the complex flows and networks of the idol-making industry uncovers the unheard voices of the residents of the neighbourhood. The emotions and the reflections of participants have been presented through the photographs they have taken. With their consent, I reflect on their emotions and respect the views presented about their neighbourhood. This research has enabled me to learn from their lived experiences and present their story to enhance the understanding of place identity within a thriving informally located cultural site. I do not present their names to keep anonymity. However, I feel the participants represent a cross section of the residents of Kumartuli.

Phase 3: February 2022

I organised a deliberative workshop on infrastructure and services in Kumartuli (Kolkata, India) on February 2022 as part of my ESRC Postdoctoral Fellowship. Deliberation is a research method that brings a small group of stakeholders with diverse views to go through a facilitated dialogue and deliberation process together. Through many years of research and fieldwork in Kolkata's Kumartuli, I identified several spatial and infrastructural concerns and grievances within the community. This small group deliberation intended to collectively reflect on these common issues and co-produce recommendations for the much-needed municipal services and infrastructure conditions in Kumartuli. The deliberative workshop discussed infrastructure, development, and inclusive planning with various experts and stakeholders. Further, it also considered how can the precarious conditions of living and working in these slum neighbourhoods be improved with their knowledge, expertise, and experience. The workshop was attended by community representatives from Kumartuli, Kolkata-based planners, architects, academic researchers, engineers, policy makers, and research students. In this workshop, community representatives such as idol-makers, *shola*-craftsmen, traders, and local residents were invited to share their experiences and grievances. The workshop aimed to provide a safe, collaborative, and fair space to discuss and share ideas for improving lives and livelihoods by enriching the infrastructure and services.

References

Agnihotri, A. (2017). *Kolkatar Pratima Shilpira*. Kolkata, Ananda Publishers Private Limited.

Bean, S.S. (2011). The unfired clay sculpture of Bengal in the artscape of modern South Asia. *A Companion to Asian Art and Architecture*. Malden, MA, Wiley Online Library: 604–628.

Cook, I. (2004). Follow the thing: Papaya. *Antipode*. **36**: 642–664.

Elder, S. (2001). Images of Asch. *Visual Anthropology Review*, Wiley Online Library. **17**: 89–109.

Ghosh, D. (2012). *Pashchimbanger Mritshilpo*. Kolkata, Ministry of Information & Cultural Affairs, Government of West Bengal.

Gubrium, A. and K. Harper (2016). *Participatory Visual and Digital Methods*. London and New York, Routledge. **10**.

Hannerz, U. (2003). Being there . . . and there . . . and there! Reflections on multi-sited ethnography. *Ethnography*. **4**: 201–216.

Harper, D. (2002). Talking about pictures: A case for photo elicitation. *Visual Studies*. **17**: 13–26.

Harvey, D. (1990). Between space and time: Reflections on the geographical imagination. *Annals of the Association of American Geographers*, Wiley Online Library. **80**: 418–434.

Heierstad, G. (2017). *Caste, Entrepreneurship and the Illusions of Tradition: Branding the Potters of Kolkata*. London, Anthem Press.

Herbert, S. (2000). For ethnography. *Progress in Human Geography*. **24**(4): 550–568.

KMDA (2009). Kumartuli Urban Renewal Project.

Knowles, C. (2014). *Flip-Flop: A Journey Through Globalisation's Backroads*. London, Pluto Press: 6–8.

Krase, J. (2012). *Seeing Cities Change: Local Culture and Class*. London, Ashgate.

Lombard, M. (2013). Using auto-photography to understand place: Reflections from research in urban informal settlements in Mexico. *Area*, Wiley Online Library. **45**(1): 23–32.

Marcus, G.E. (1995). Ethnography in/of the world system: The emergence of multi-sited ethnography. *Annual Review of Anthropology*. **24**: 95–117.

Margolis, E. (1988). Mining photographs: Unearthing the meanings of historical photos. *Radical History Review*. **1988**: 33–49.

Margolis, E. (1994). Images in struggle: Photographs of Colorado coal camps. *Visual Studies*. **9**: 4–26.

Milne, E.J. (2012). *Handbook of Participatory Video—9780759121133*. Lanham, MD, Altamira Press.

Moser, C. and A. Stein (2011). Implementing urban participatory climate change adaptation appraisals: A methodological guideline. *Environment and Urbanisation*. **23**: 463–485.

Mylan, J. and D. Southerton (2017). *Following the Action: An Approach for Studying the Coordination of Practice—Practice Theory Methodologies*. Retrieved June 7, 2023, from https://practicetheorymethodologies.wordpress.com/2017/03/22/jo-mylan-dale-souther ton-following-the-action-an-approach-for-studying-the-coordination-of-practice/.

Prosser, J. and A. Loxely (2008). ESRC National Centre for Research Methods Review Paper: Introducing visual methods. *National Centre for Research Methods Review Paper NCRM/010*: 1–65.

Schwartz, D. (1989). Visual ethnography: Using photography in qualitative research. *Qualitative Sociology*. **12**: 119–154.

Shove, E. (2017). *Practice Theory Methodologies Do Not Exist—Practice Theory Methodologies*. Retrieved June 7, 2023, from https://practicetheorymethodologies.wordpress.com/2017/02/15/elizabeth-shove-practice-theory-methodologies-do-not-exist/.

Wang, C. and M.A. Burris (1994). Empowerment through photo novella: Portraits of participation. *Health Education & Behavior*. **21**: 171–186.

Wigglesworth, S. and J. Till (1998). The everyday and architecture. *Architectural Design*. **134**.

Wilson, S. and E.J. Milne (2016). Visual activism and social justice: Using visual methods to make young people's complex lives visible across 'public' and 'private' spaces. *Current Sociology*. **64**: 140–156.

Glossary of Bengali words

auto-rickshaws   A mode of transit in inner-city areas; three-wheeled motorised vehicles

babu   An aristocratic, well-dressed man of the emerging Bengali wealthy merchant class in the 1800 and 1900s. Also, used to address Bengali men.

baiji   A professional woman singer and dancer trained in Hindustani Classical music

bangalpara   A *para* dominated by people of Bangladeshi origin

Basanti   Another female deity, associated with Durga; the colour of Spring—yellow

bari   Home

baro   Twelve

baroyari/barowari   Participated by all

basti/bustee   A colloquial term used widely in India for a settlement

bazaar   A market

bele-mati   Sandy clay

bhadrolok   Gentleman: a term used to refer to men of the privileged classes

Boishakh   The first month of the Bengali calendar, during April–May

bonedi bari   Elite mansions

Brahmin   Priest, priest caste

daker saj   Metallic ornaments

dhaak   A drum

dhaki/dhaaki   A drummer

diya   A lamp

entel-mati   Sticky clay required for sculpting

ganga-mati   Clay from the bed of the river Ganges (in this case, river Hooghly)

ghat   Riverfront embankment

jali   Mesh or lattice-like configuration of ventilators prevalent in South Asian building construction techniques

jamidar/zamindar   Landlords or landholding families

jati   Sometimes translates as caste system (occupational)

jatrapala/jatra   Bengali equivalent of opera

Kailash   A mountain range imagined somewhere in the higher snow-capped reaches of the Himalayas

kala-mati Darker clay

karmakar/kamar A metal worker

kathamo A frame, structure of the idol, made of bamboo and wood

Kayastha A caste, prevalent mainly in Bengal

khari-mati White clay

Kshatriya Warrior caste

kumar Potter

kutcha/kacha Provisional, temporary; also widely used to mean unfinished wet idol

Lakshmi, Saraswati, Ganesh, and Kartik Four mythological children of Durga. Each of the four children represents individual Hindu deities

macha The mezzanine floor or temporary structures made of wooden slats or bamboo

madhyabitta Middle class

mahajan A merchant or supplier, sometimes the local banker who provides a loan

mahisasura The demon deity who is associated with the buffalo.

mahisasur-mardini The female deity who is the *Mahisasura* demon slayer

majhergoli Middle alley, an alley in the Kumartuli neighbourhood

malakar/mali A person belonging to the florist and garland-weaving caste

malik Owner of the business, master

mati Clay

modak/moyra A confectioner

mritshilpi Artists who work with clay

munshi Teacher of local language and trade

nouka A boat

pandals Make-shift pavilions to house the Durga idols

para Loosely translated as a neighbourhood or locality, but has various meanings

paribartan Change, reform

pheriwala A mobile street vendor

pukur A pond

pukka/paka Permanent

rajbari Mansion of the king

rathayatra A festival of the chariot celebrated during monsoon, usually in June/July

Sabeki Traditional

saj Ornaments or harness

sarani A wide road

sarbojonin For everyone to participate, all-inclusive

sari An outfit consisting of a long piece of cloth wrapped around the body, primarily worn by women from South Asia

shakti Power or energy

shola dried milky-white spongey plant matter

Sudra Working caste, even menial

swadesi of or belonging to one's country, often referred to as India's freedom fighters' movement in Bengal during the colonial period

tanti/tantubay   A weaver

terracotta   Fired clay

thakurpotti   Neighbourhood of idol-makers

thana   Police station

Thika   Lease-hold

tola/tuli   *Tola* is an area of a homogeneous population group; *tuli* is a smaller *tola*

tubri, uron-tubri   *Tubri* is a spherical hollow firework stuffed with charcoal and gunpowder.

Uron-tubri   is now widely banned for fire-safety and was a flying variant of *tubri*

Vaishya or Baniya   Merchants

Varna   Caste system (social)

yar   A friend

zari   Alternate material for ornaments, and a thread of lace looks similar to gold or silver strips

Index

Note: Page numbers in *italics* indicate a figure and page numbers in **bold** indicate a table on the corresponding page.

Taylor & Francis Group
an **informa** business

Taylor & Francis eBooks

www.taylorfrancis.com

A single destination for eBooks from Taylor & Francis
with increased functionality and an improved user
experience to meet the needs of our customers.

90,000+ eBooks of award-winning academic content in
Humanities, Social Science, Science, Technology, Engineering,
and Medical written by a global network of editors and authors.

TAYLOR & FRANCIS EBOOKS OFFERS:

A streamlined
experience for
our library
customers

A single point
of discovery
for all of our
eBook content

Improved
search and
discovery of
content at both
book and
chapter level

REQUEST A FREE TRIAL
support@taylorfrancis.com

Routledge
Taylor & Francis Group

CRC Press
Taylor & Francis Group

For Product Safety Concerns and Information please contact our EU
representative GPSR@taylorandfrancis.com
Taylor & Francis Verlag GmbH, Kaufingerstraße 24, 80331 München, Germany

www.ingramcontent.com/pod-product-compliance
Lightning Source LLC
Chambersburg PA
CBHW060307220326
41598CB00027B/4262

9 78 1 0 3 2 3 7 6 5 0 9